ENCYCLOPAEDIA OF CLASSIFICATION OF FISH

ENCYCLOPAEDIA OF CLASSIFICATION OF FISH

Vol. 1

A.M. Bagulia

ANMOL PUBLICATIONS PVT. LTD.

NEW DELHI - 110 002 (INDIA)

ANMOL PUBLICATIONS PVT. LTD.
H.O.: 4374/4B, Ansari Road, Darya Ganj,
New Delhi-110 002 (India)
Ph.: 23278000, 23261597
B.O.: No. 1015, Ist Main Road, BSK IIIrd Stage
IIIrd Phase, IIIrd Block,
Bangalore - 560 085 (India)
Visit us at: www.anmolpublications.com

Encyclopaedia of Classification of Fish

First Published, 2008

ISBN 978-81-261-3530-1 (Set)

PRINTED IN INDIA

Printed at Mehra Offset Press, Delhi.

Contents

Preface

Fish are classified in several classes by which the study of fish becomes easier. Fish having similar properties are contained in the same group. Various classes of fish are Thelodonti, Anaspida, Cephalaspidomorphi, Galeaspida, Pituriaspida, Osteostracy, Gnathostomata, Placodermi. Jawless fish are contained under the group Thelodonti, which are very similar to Heterostraci, and are not armoured.

The Pituriaspida are a small group of armoured jawless fish with tremendous nose-like rostrums, which live in the marine, deltaic environments of Middle Devonian Australia. They are known only by two species, *Pituriaspis doylei* and *Neeyambaspis enigmatica,* found in a single sandstone location of the Georgina Basin, in Western Queensland, Australia. The Osteostraci is a group of bony-armoured jawless fish, which lived in North America, Europe and Russia from the Middle Silurian to Late Devonian Period. The Devonian species were among the most advanced of all known Agnathans. This was due to development of paired fins, and their complicated cranial anatomy. However, the osteostracans were more related to other Aghathans then to jawed ventebrates as their inner ears were formed by two pairs of semi-circular canals.

Gnathostomata is the group of vertebrates with jaws. This group is in superclass, including the familiar classes of fish, birds, mammals, and so forth and a sister group of jawless vertebrates Agnatha. The Placodermi are armoured prehistoric fish known from fossils, dating from the Late Silurian to the end of the Devonian Periods. Their head and thorax were covered by articulated armoured plates and the rest of the body was either scaled or naked. Placoderms were among the first of the jawed fish, their jaws likely to have evolved from the first of their gill arches. The first identifiable Placoders evolved in the Late Silurian; they disappeared in the Late Devonian extinctions. Thus, the Fish world is a world in itself.

This eminent work has an exhaustive and exclusive coverage of fish in a simple, but effective style. Let's hope, this work would be capable to prove to be beneficial for all classes of readers.

Editor

1

Righteye Flornders and River Stingrays

Pleuronectidae

Righteye flounders are a family, Pleuronectidae, of flounders. They are called "righteye flounders" because most species lie on the sea bottom on their left side, with both eyes on the right side.

Their dorsal and anal fins are long and continuous, with the dorsal fin extending forward onto the head.

They are found on the bottoms of oceans around the world, with some species, such as the Atlantic halibut, *Hippoglossus hippoglossus*, being found down to 2,000 m. The smaller species eat sea-floor invertebrates such as polychaetes and crustaceans, but the larger righteye flounders, such as *H. hippoglossus*, which grows up to 2.4 m, feed on other fishes and cephalopods as well.

They include many important commercially fished species, including not only the various fish called flounders, but also the European plaice, the halibuts, the lemon sole, the common dab, the Pacific Dover sole, and the flukes.

Classification

In some classifications the subfamilies Paralichthodinae, Poecilopsettinae, and Rhombosoleinae are raised to the level of families.

According to FishBase there are 101 species in 41 genera and five subfamilies:

- Subfamily Eopsettinae
  - — Genus *Atheresthes*
    - ⇒ Kamchatka flounder, *Atheresthes evermanni* (Jordan & Starks, 1904).
    - ⇒ Arrowtooth flounder, *Atheresthes stomias* (Jordan & Gilbert, 1880).
  - — Genus *Eopsetta*
    - ⇒ Shotted halibut, *Eopsetta grigorjewi* (Herzenstein, 1890).
    - ⇒ Petrale sole, *Eopsetta jordani* (Lockington, 1879).
- Subfamily Hippoglossinae
  - — Genus *Clidoderma*
    - ⇒ Roughscale sole, *Clidoderma asperrimum* (Temminck & Schlegel, 1846).
  - — Genus *Hippoglossus*
    - ⇒ Atlantic halibut, *Hippoglossus hippoglossus* (Linnaeus, 1758).
    - ⇒ Pacific halibut, *Hippoglossus stenolepis* (Schmidt, 1904).
  - — Genus *Reinhardtius*
    - ⇒ Greenland halibut, *Reinhardtius hippoglossoides* (Walbaum, 1792)
  - — Genus *Verasper*
    - ⇒ Barfin flounder, *Verasper moseri* (Jordan & Gilbert, 1898).
    - ⇒ Spotted halibut, *Verasper variegatus* (Temminck & Schlegel, 1846).

- Subfamily Hippoglossoidinae
  - — Genus *Acanthopsetta*
    - ⇒ Scale-eye plaice, *Acanthopsetta nadeshnyi* (Schmidt, 1904).
  - — Genus *Cleisthenes*
    - ⇒ Sohachi, *Cleisthenes herzensteini* (Schmidt, 1904).
    - ⇒ *Cleisthenes pinetorum* (Jordan & Starks, 1904).
  - — Genus *Hippoglossoides*
    - ⇒ Flathead flounder, *Hippoglossoides dubius* (Schmidt, 1904).
    - ⇒ Flathead sole, *Hippoglossoides elassodon* (Jordan & Gilbert, 1880).
    - ⇒ American plaice, *Hippoglossoides platessoides* (Fabricius, 1780)
    - ⇒ Bering flounder, *Hippoglossoides robustus* (Gill & Townsend, 1897).
- Subfamily Lyopsettinae
  - — Genus *Lyopsetta*
    - ⇒ Slender sole, *Lyopsetta exilis* (Jordan & Gilbert, 1880).
- Subfamily Paralichthodinae
  - — Genus *Paralichthodes*
    - ⇒ Peppered flounder, *Paralichthodes algoensis* (Gilchrist, 1902).
- Subfamily Pleuronectinae
  - — Tribe Isopsettini
    - ⇒ Genus *Isopsetta*
    - ⇒ Butter sole, *Isopsetta isolepis* (Lockington, 1880).
  - — Tribe Microstomini
    - ⇒ Genus *Dexistes*
    - ⇒ Rikuzen flounder, *Dexistes rikuzenius* (Jordan & Starks, 1904).
    - ⇒ Genus *Embassichthys*

⇒ Deepsea sole, *Embassichthys bathybius* (Gilbert, 1890).

⇒ Genus *Glyptocephalus*

⇒ Witch, *Glyptocephalus cynoglossus* (Linnaeus, 1758)

⇒ Blackfin flounder, *Glyptocephalus stelleri* (Schmidt, 1904).

⇒ Rex sole, *Glyptocephalus zachirus* (Lockington, 1879).

⇒ Genus *Hypsopsetta*

⇒ Diamond turbot, *Hypsopsetta guttulata* (Girard, 1856).

⇒ *Hypsopsetta macrocephala* (Breder, 1936).

⇒ Genus *Lepidopsetta*

⇒ Rock sole, *Lepidopsetta bilineata* (Ayres, 1855).

⇒ Dusky sole, *Lepidopsetta mochigarei* (Snyder, 1911).

⇒ Northern rock sole, *Lepidopsetta polyxystra* (Orr & Matarese, 2000).

⇒ Genus *Microstomus*

⇒ Slime flounder, *Microstomus achne* (Jordan & Starks, 1904).

⇒ Lemon sole, *Microstomus kitt* (Walbaum, 1792)

⇒ Dover sole, *Microstomus pacificus* (Lockington, 1879).

⇒ *Microstomus shuntovi* (Borets, 1983).

⇒ Genus *Pleuronichthys*

⇒ C-O sole, *Pleuronichthys coenosus* (Girard, 1854).

⇒ Ridged-eye flounder, *Pleuronichthys cornutus* (Temminck & Schlegel, 1846).

⇒ Curlfin sole, *Pleuronichthys decurrens* (Jordan & Gilbert, 1881).

⇒ Ocellated turbot, *Pleuronichthys ocellatus* (Starks & Thompson, 1910).

⇒ Spotted turbot, *Pleuronichthys ritteri* (Starks & Morris, 1907).

⇒ Hornyhead turbot, *Pleuronichthys verticalis* (Jordan & Gilbert, 1880).

⇒ Genus *Tanakius*

⇒ Willowy flounder, *Tanakius kitaharae* (Jordan & Starks, 1904).

— Tribe Pleuronectini

⇒ Genus *Kareius*

⇒ Stone flounder, *Kareius bicoloratus* (Basilewsky, 1855).

⇒ Genus *Limanda*

⇒ Yellowfin sole, *Limanda aspera* (Pallas, 1814).

⇒ Yellowtail flounder, *Limanda ferruginea* (Storer, 1839).

⇒ Common dab, *Limanda limanda* (Linnaeus, 1758)

⇒ Longhead dab, *Limanda proboscidea* (Gilbert, 1896).

⇒ Sand flounder, *Limanda punctatissimus* (Steindachner, 1879).

⇒ Sakhalin sole, *Limanda sakhalinensis* (Hubbs, 1915).

⇒ Genus *Liopsetta*

⇒ Arctic flounder, *Liopsetta glacialis* (Pallas, 1776).

⇒ Far Eastern smooth flounder, *Liopsetta pinnifasciata* (Norman, 1926).

⇒ American smooth flounder, *Liopsetta putnami* (Gill, 1864).

⇒ Genus *Parophrys*

⇒ English sole, *Parophrys vetulus* (Girard, 1854).

⇒ Genus *Platichthys*

⇒ European flounder, *Platichthys flesus* (Linnaeus, 1758).

⇒ Starry flounder, *Platichthys stellatus* (Pallas, 1788).

⇒ Genus *Pleuronectes*

⇒ European plaice, *Pleuronectes platessa* (Linnaeus, 1758).

⇒ Alaska plaice, *Pleuronectes quadrituberculatus* (Pallas, 1814).

⇒ Genus *Pseudopleuronectes*

⇒ Winter flounder, *Pseudopleuronectes americanus* (Gill, 1864).

⇒ Littlemouth flounder, *Pseudopleuronectes herzensteini* (Schmidt, 1904).

⇒ *Pseudopleuronectes obscurus* (Herzenstein, 1890).

⇒ Cresthead flounder, *Pseudopleuronectes schrenki* (Schmidt, 1904).

⇒ Marbled flounder, *Pseudopleuronectes yokohamae* (Gunther, 1877).

— Tribe Psettichthyini

⇒ Genus *Psettichthys*

⇒ Pacific sand sole, *Psettichthys melanostictus* (Girard, 1854).

• Subfamily Poecilopsettinae

— Genus *Marleyella*

⇒ Comb flounder, *Marleyella bicolorata* (Basilewsky, 1855).

⇒ *Marleyella maldivensis* (Norman, 1939).

— Genus *Nematops*

⇒ Narrow-body righteye flounder, *Nematops chui* (Fowler, 1934).

⇒ Largescale righteye flounder, *Nematops grandisquama* (Weber & de Beaufort, 1929).

⇒ Long-fin righteye flounder, *Nematops macrochirus* (Norman, 1931).

⇒ Small-mouth righteye flounder, *Nematops microstoma* (Gunther, 1880).

⇒ *Nematops nanosquama* (Amaoka, Kawai & Seret, 2006).

— Genus *Poecilopsetta*

⇒ *Poecilopsetta albomaculata* (Norman, 1939).

⇒ Deepwater dab, *Poecilopsetta beanii* (Goode, 1881)

⇒ Coloured righteye flounder, *Poecilopsetta colorata* (Gunther, 1880).

⇒ *Poecilopsetta dorsialta* (Guibord & Chapleau, 2001).

⇒ *Poecilopsetta hawaiiensis* (Gilbert, 1905).

⇒ *Poecilopsetta inermis* (Breder, 1927).

⇒ *Poecilopsetta macrocephala* (Breder, 1936).

⇒ Fowler's large-scale righteye flounder, *Poecilopsetta megalepis* (Fowler, 1934).

⇒ African righteye flounder, *Poecilopsetta natalensis* (Norman, 1931).

⇒ *Poecilopsetta normani* (Foroshchuk & Fedorov, 1992).

⇒ *Poecilopsetta pectoralis* (Kawai & Amaoka, 2006).

⇒ Tile-coloured righteye flounder, *Poecilopsetta plinthus* (Jordan & Starks, 1904).

⇒ Alcock's narrow-body righteye flounder, *Poecilopsetta praelonga* (Alcock, 1894).

⇒ *Poecilopsetta vaynei* (Quero, Hensley & Mauge, 19880.

⇒ *Poecilopsetta zanzibarensis* (Norman, 1939).

- Subfamily Rhombosoleinae

— Genus *Ammotretis*

⇒ Shortfin flounder, *Ammotretis brevipinnis* (Norman, 1926).

⇒ Elongate flounder, *Ammotretis elongatus* (Yarrell, 1839).

⇒ Tudor's flounder, *Ammotretis lituratus* (Richardson, 1844).

⇒ *Ammotretis macrolepis* (McCulloch, 1914).

⇒ Longsnout flounder, *Ammotretis rostratus* (Gunther, 1862).

— Genus *Azygopus*

⇒ Banded-fin flounder, *Azygopus pinnifasciatus* (Norman, 1926).

— Genus *Colistium*

⇒ New Zealand brill, *Colistium guntheri* (Hutton, 1873).

⇒ New Zealand turbot, *Colistium nudipinnis* (Waite, 1911).

— Genus *Oncopterus*

⇒ Remo flounder, *Oncopterus darwinii* (Steindachner, 1874).

— Genus *Pelotretis*

⇒ Southern lemon sole, *Pelotretis flavilatus* (Waite, 1911).

— Genus *Peltorhamphus*

⇒ Speckled sole, *Peltorhamphus latus* (James, 1972).

⇒ New Zealand sole, *Peltorhamphus novaezeelandiae* (Gunther, 1862).

⇒ *Peltorhamphus tenuis* (James, 1972).

— Genus *Psammodiscus*

⇒ Indonesian ocellated flounder, *Psammodiscus ocellatus* (Starks & Thompson, 1910).

— Genus *Rhombosolea*

⇒ Yellowbelly flounder, *Rhombosolea leporina* (Gunther, 1862).

⇒ New Zealand flounder, *Rhombosolea plebeia* (Richardson, 1843).

⇒ Black flounder, *Rhombosolea retiaria* (Hutton, 1873).

⇒ Greenback flounder, *Rhombosolea tapirina* (Gunther, 1862).

— Genus *Taratretis*

⇒ Derwent flounder, *Taratretis derwentensis* (Last, 1978).

River Shark

The river sharks are six rare species of shark in the genus *Glyphis*, although, due to their secretive habits, other species could easily remain undiscovered. The river sharks are members of the family Carcharhinidae, and thus share the basic characteristics of the group. The bull shark is sometimes called both the river shark and the Ganges shark; it should not be confused with the true river sharks of *Glyphis*.

Characteristics

In general, all river sharks feature the following field characteristics:

- A short, broadly rounded snout (its length less than the mouth width);
- Small, widely spaced nostrils;
- Small, dark eyes;
- Broad, serrated upper teeth;
- Very short labial furrows (lip grooves), restricted to the jaw corners;
- A broad dorsal fin with the mid-base closer to the base of the pectoral fins than those of the pelvic fins; and
- An anal fin with a deeply excised posterior margin.

Morphology

River sharks are very similar in overall morphology to requiem sharks of the genus Carcharhinus, but can be distinguished from them by the following characteristics:

- Cusps of lower teeth protrude prominently when mouth is closed;
- The second dorsal fin is 1/2 to 3/5 the height of the first dorsal fin;
- The origin of the second dorsal fin is slightly anterior to the origin of the anal fin;
- Precaudal pit is longitudinal rather than crescent-shaped.

Species

- Ganges shark, *Glyphis gangeticus* (Muller & Henle, 1839).
- Speartooth shark, *Glyphis glyphis* (Muller & Henle, 1839).
- Irrawaddy River shark, *Glyphis siamensis* (Steindachner, 1896).
- Bizant river shark, *Glyphis sp. A*, not yet described.

- Borneo river shark, *Glyphis sp. B*, not yet described.
- New Guinea river shark, *Glyphis sp. C*, not yet described.

Potamotrygonidae

River stingrays are neotropical freshwater fishes of the Potamotrygonidae family (order Rajiformes).

They are native to eastern South America, living in rivers that drain into the Caribbean, and into the Atlantic as far south as the Rio de la Plata in Argentina. Generally, each species is native to a single river basin.

River stingrays are almost circular in shape, and range in size from *Potamotrygon schuhmacheri*, which reaches 25 cm in diameter, to the ocellate river stingray, *Potamotrygon motoro*, which grows up to a metre in diameter. The dorsal surface is covered with denticles (sharp tooth-like scales).

They have a poisonous caudal sting, and are one of the most feared freshwater fishes in the neotropical region, sometimes more feared than piranhas and electric eels. However, they are not dangerous unless stepped on or otherwise threatened.

River stingrays are the only family of batoids to be restricted to fresh water habitats.

Species

There are nineteen species in four genera:

- Genus *Paratrygon*
  - Discus ray, *Paratrygon aiereba* (Muller & Henle, 1841).
- Genus *Plesiotrygon*
  - Long-tailed river stingray, *Plesiotrygon iwamae* (Rosa, Castello & Thorson, 1987).
- Genus *Potamotrygon*
  - Short-tailed river stingray, *Potamotrygon brachyur* (Gunther, 1880).
  - Vermiculate river stingray, *Potamotrygon castexi* (Castello & Yagolkowski, 1969).

- — Thorny river stingray, *Potamotrygon constellata* (Vaillant, 1880).
- — Largespot river stingray, *Potamotrygon falkneri* (Castex & Maciel, 1963).
- — Bigtooth river stingray, *Potamotrygon henlei* (Castelnau, 1855).
- — Porcupine river stingray, *Potamotrygon hystrix* (Muller & Henle, 1834).
- — White-blotched river stingray, *Potamotrygon leopoldi* (Castex & Castello, 1970).
- — Magdalena river stingray, *Potamotrygon magdalenae* (Dumeril, 1865).
- — Ocellate river stingray, *Potamotrygon motoro* (Muller & Henle, 1841).
- — Red-blotched river stingray, *Potamotrygon ocellata* (Engelhardt, 1912).
- — Smooth back river stingray, *Potamotrygon orbignyi* (Gunther, 1880).
- — Rosette river stingray, *Potamotrygon schroederi* (Fernandez-Yepez, 1957).
- — *Potamotrygon schuhmacheri* (Castex, 1964).
- — Raspy river stingray, *Potamotrygon scobina* (Garman, 1913).
- — Parnaiba river stingray, *Potamotrygon signata* (Garman, 1913).
- — Maracaibo river stingray, *Potamotrygon yepezi* (Castex & Castello, 1970).

- Genus *Trygon*
  - — *Trygon garrapa* Jardine, 1843.

Roanoke Bass

The Roanoke bass (*Ambloplites cavifrons*) is a species of freshwater fish in the sunfish family (Centrarchidae) of order Perciformes. It is native only to a few river systems (including the Roanoke) in Virginia and North Carolina.

A. cavifrons reaches a maximum recorded overall length of 36 cm (14 in) and a maximum recorded weight of 620 g (1.4 lb).

It prefers clear, rocky creeks and pools.

The Roanoke bass is considered Vulnerable because its very small range renders it sensitive to danger from human activity.

Rock Bass

The rock bass (*Ambloplites rupestris, Ambloplites Ariommus, Ambloplites Constellatus*), also known as the *rock perch* or *goggle-eye,* is a species of freshwater fish in the sunfish family (Centrarchidae) of order Perciformes.

Rock bass are native to the St. Lawrence River and Great Lakes system, the upper and middle Mississippi River basin in North America from Quebec to Saskatchewan in the north down to Missouri and Arkansas, and throughout the eastern US from New York through Kentucky and Tennessee to the northern portions of Alabama and Georgia in the south. All species of rock bass are considered gamefish and are popular with sustenance and sport fishermen. Sport anglers often employ ultra-light spinning gear or fly tackle designed for panfish.

A. rupestris, the largest and most common of the *Ambloplites* species, has reached a maximum recorded length of 43 cm (17 in.), and a maximum recorded weight of 1.4 kg (3.0 lb). It can live as long as 10 years. These fish have the ability to rapidly change their colour to match their surroundings. It is this chameleon-like trait that allows them to thrive throughout their wide range.

The rock bass prefers clear, rocky, and vegetated stream pools and lake margins. It is carnivorous, and its diet consists of smaller fish, insects, and crustaceans.

Ambloplites Constellatus, a species of rock bass from the Ozark upland of Arkansas, and *Ambloplites ariommus* are true rock bass, but regarded as separate species.

A. rupestris is sometimes called the *redeye* or *redeye bass* in Canada, but this name refers more properly to *Micropterus coosae,* a distinct species of Centrarchid native to parts of the American South. Rafinesque originally assigned the rock bass to *Bodianus,* a genus of marine wrasses (family Labridae).

Rock Beauty

The rock beauty, *Holacanthus tricolor*, is a species of marine angelfish of the family Pomacanthidae, found in the western Atlantic from Georgia, United States, Bermuda, and the northern Gulf of Mexico to Rio de Janeiro, Brazil, at depths of between 3 and 92 m. Its length is up to 35 cm.

The rock beauty is inhabits rock jetties, rocky reefs and rich coral areas. Juveniles are often associated with fire corals. It feeds on tunicates, sponges, zoantharians and algae.

Colouration of the front of the body is yellow; the remaining parts of body, dorsal fin, and the front of anal fin are black. The caudal fin is entirely yellow. The front margin of the anal fin and edge of the gill cover are orange; bright blue on the upper and lower part of the iris. The young of about an inch in length are entirely yellow except for a blue-edged black spot on the upper side of the body posterior to the mid-point; with growth, the black spot soon expands to become the large black area covering most of the body and dorsal and anal fins.

Rock Cod

The rock cod (*Lotella rhacina*) is a temperate fish found off the coasts of southeastern Australia, Tasmania, the Great Australian Bight and northwards up the south western Australia coasts. They are also found around the coasts of New Zealand. They belong to the family Gadidae and are thus related to the true cods (genus *Gadus*). They are also known as beardie in Australia.

Rock cod are yellow-grey to red-brown with white fin margins. They have chin barbels. They may grow up to 50cm in length. They are found in caves in bays and coastal reefs.

They are frequently found inshore and inhabit shallow waters in the continental shelf with typical depth of 10 to 90 metres.

Note that there are many other fish that are sometimes referred to as rock cod, but most of these are unrelated to the cod family, and are better known as groupers.

Yoma Danio

The Yoma Danio (*Danio feegradei*) is a fish in the Cyprinid family, a species of Danio from Burma. Although discovered by Hora in 1937 it was first exported in 2005.

This fish has an exceptional ability for jumping. The fish can jump vertically to a height of over a foot (0.3 m) and it is not unknown to jump in this manner repeatedly. A tight fitting lid with no gaps is recommended.

The Yoma danio is quite variable with some specimens developing black bars.

- Maximum length: 3 in (8 cm)
- Colours: blue, gold, black, red
- Temperature preference: 20-25 degrees Celsius
- pH preference: 6 to 7
- Hardness preference: Soft to medium
- Salinity preference: Low to medium
- Compatibility: Good but fast like most danios, a largeish fish though, needs plenty of space
- Life span: Typically 3 to 5 years
- Ease of keeping: Moderate
- Ease of breeding: Moderate to hard

Rockfish

A striped bass. Used most frequently in the mid-eastern states where the fish spawns (Delaware, Maryland and Virginia). Rockfish may refer to one of the following fishes:

- Striped bass, a member of the Moronidae (common name: temperate bass) family

- A member of the Sebastidae family of venomous fish of the Scorpaeniformes order
- The New Zealand rockfish (Acanthoclinus littoreus)
- Sebastes (Alaskan rockfish) estimated to live up to 157 years, weighing approximately 60lbs, 40"-50" in length & of an orange colour, these fish live approximately 2000' below the surface of the water. One such fish was caught off the Anchorage, Alaska coast in March of 2007(GMC-KUA).

Rockfish may also refer to:

- Rockfish Seafood Grill, a Dallas, Texas restaurant chain.
- Rockfish River, a river in the Blue Ridge Mountains in Virginia.
- Rockfish Interactive, an Arkansas-based digital marketing and technology agency.
- *Rockfish* (short film), a 2002 computer-animated short by Tim Miller of Blur Studio.
- *Rockfish* (film), a full-length film based on the short, being co-produced by and starring Vin Diesel for a release in 2007.

Rohu

Rohu (*Labeo rohita*) is a fish of the carp family Cyprinidae, found commonly in rivers and freshwater lakes in and around the South Asia and South East Asia. It is also known as *rawas* in Hindi, *rui* in Bengali, *rou* in Assamese and is popular in Thailand, Pakistan, Bangladesh, Orissa, West Bengal, Assam, and the Konkan region of India. It is a non-oily/white fish.

The roe of rohu is also considered as a delicacy by Bengalis. It is deep fried and served hot as an appetiser as part of a Bengali meal. It is also stuffed inside pointed gourd to make *potoler dorma* which is a delicacy often prepared to satisfy the palate of the discerning guest.

Rohu is also served deep fried in mustard oil, as *kalia* which is a rich gravy made of concoction of spices and deeply browned onions and *tok*, where the fish is cooked in a flavourful and tangy sauce made of tamarind and mustard. Rohu is also very popular in Northern India such as in the province of Punjab. It is a speciality of Lahori cuisine as in Lahori fried fish prepared with batter and spices.

The Rohu is also a famed sportfish and gives an adrenaline charged fight, often leaping several feet into the air.

Ronquil

Ronquils (sometimes spelt ronchils) are perciform marine fish of the small family Bathymasteridae. Found only in Arctic and North Pacific waters, the ronquil family contains just seven species in three genera. The larger species are important to commercial fisheries as food fish. Ronquils are most closely related to the eelpouts and prowfish.

The name *ronquil* is said to derive from the Spanish *ronquillo* meaning "slightly hoarse". The family name *Bathymasteridae* can be translated from the Greek to mean "deep searcher".

Physical Description

With predominantly drab colours in shades of orange and olive, greatly elongate bodies and large rounded heads, ronquils could easily be mistaken for blennies. Their eyes and terminal mouths are large, with a single long, continuous dorsal fin which begins just behind the head; the anal fin is similarly extensive, and both it and the dorsal fin may either have 1-2 spines, or none at all. The pectoral fins are also quite large, and like the much smaller tail fin, rounded. The reduced pelvic fins are located in a thoracic position, just behind the throat.

The lateral line runs high along the flanks of the ronquil, ending with the dorsal fin. The fish have *palatine teeth*; that is, teeth located on the palate or roof of the mouth. The largest of the seven species is the searcher (*Bathymaster signatus*) at up to 30.5 centimetres in length.

Marked sexual dimorphism is observed among ronquils. For example, male northern ronquil (*Ronquilus jordani*) have orange backs with iridescent blue anal fins bordered in black, while the females have olive backs and paler blue anal fins bordered in brown.

Habitat and Behaviour

Inhabiting cold waters from the Siberian and Bering Seas south to Hokkaido and Monterey, California, ronquils are benthic animals spending most of their time on or near the bottom. Ronquils may be found on sandy or rocky substrates, usually no deeper than 275 metres (although searchers have been recorded to 825 metres); some may be found in tide pools. When threatened, ronquils will choose to retreat into the nearest hiding place, such as a hole or crevice.

Ronquils feed primarily on small benthic crustaceans and molluscs. Large halibut and flounders are among the most significant predators of ronquils. Although considered fairly common, ronquils are rarely observed; this is likely owing to their secretive nature.

Few specifics are recorded regarding ronquil spawning, but the males are known to guard the brood.

Species

- Genus: *Bathymaster*
  - — *Bathymaster caeruleofasciatus*
  - — *Bathymaster derjugini*
  - — *Bathymaster leurolepis*
  - — *Bathymaster signatus*
- Genus: *Rathbunella*
  - — *Rathbunella alleni*
  - — *Rathbunella hypoplecta*
- Genus: *Ronquilus*
  - — *Ronquilus jordani*

Roosterfish

The roosterfish, *Nematistius pectoralis*, is a game fish common in the marine waters surrounding Mexico, from the Gulf of

California to Panama, and in the eastern Pacific, from California to Peru. It is the only fish in the genus *Nematistius* and the family Nematistiidae. It is distinguished by its "rooster comb", seven very long spines of the dorsal fin.

The roosterfish has an unusual arrangement of its ears: the swim bladder penetrates the brain through the large foramina and makes contact with the inner ear. It uses its swim bladder to amplify sounds. Rooster fish can reach 4 feet in length and over one hundred pounds. The weight of the average fish hooked is about 20 pounds. The fish is popular as a game fish and is also excellent to eat.

Reedfish

The reedfish, *Erpetoichthys calabaricus*, ropefish, or snakefish is a species of freshwater fish in the bichir family and order. It is the only member of the genus *Erpetoichthys*. It is native to West Africa, with its natural habitat stretching from Nigeria to the Congo. The reedfish has a maximum total length of 45 centimetres (17.72 inches). It lives in slow-moving, brackish, warm water, and it can breathe air (meaning it is able to survive in water with low dissolved oxygen content). The reedfish is a nocturnal creature, feeding on annelid worms, crustaceans and insects at night, and it is sometimes displayed in aquariums. Its genus name *Erpetoichthys* derives from the Greek words *erpeton* ("creeping thing") and *ichthys* ("fish").

In the Aquarium

Although reedfish are not commonly found in aquarium stores, they make wonderful pets. They are generally priced from $20 to about $40, however, as no recorded breeding in captivity has been seen, they must be imported from Africa, meaning their price is unlikely to decrease. They are inquisitive, peaceful, and have some "personality." Although nocturnal, reedfish will sometimes come out during the day, and this can be encouraged by daytime feeding of bloodworms or, for larger fish, nightcrawlers. The fish are known to jump, so the aquarium should have a tightly-fitting lid without large holes.

Slimehead

Slimeheads, also known as roughies and redfish, are mostly small, exceptionally long-lived, deep-sea beryciform fish comprising the family Trachichthyidae (derived from the Greek *trachys* ["rough"] and *ichthys* ["fish"]). Found in temperate to tropical waters of the Atlantic, Indian, and Pacific Ocean, the family contains approximately 45 species in eight genera. Slimeheads are named for the network of muciferous canals riddling their heads.

The larger species — namely the orange roughy (*Hoplostethus atlanticus*) and Darwin's slimehead (*Gephyroberyx darwinii*) — are the target of extensive commercial fisheries off Australia and New Zealand. Many populations have already crashed, while others are showing signs of severe overfishing; due to slimeheads' slow rate of reproduction, the future viability of these fisheries has been put into question. Orange roughies are food fish and are marketed fresh and frozen, whereas Darwin's slimeheads are utilised for their oil and made into fishmeal.

Physical Description

With a typically deep-bodied, laterally compressed form, slimeheads are conspicuous for their large, titular heads, large eyes, and (in some species) bright colours. The head is especially notable for its network of mucus-filled canals, which constitute the cranial portion of the lateral line system.

Similar cranial networks are found in the beryciform fangtooths (Anoplogastridae) and the stephanoberyciform ridgeheads (Melamphaidae). The trachichthyid head is typically blunt with a large and oblique mouth; the snout may project slighty in front of the upper jaw.

A short, sharp spine is present on the preoperculum and/or operculum and post-temporal bone, the latter spine directed posteriorly. Species of the genera *Optivus*, *Paratrachichthys*, and *Sorosichthys* differ in form from other members of the family; their bodies are more elongate.

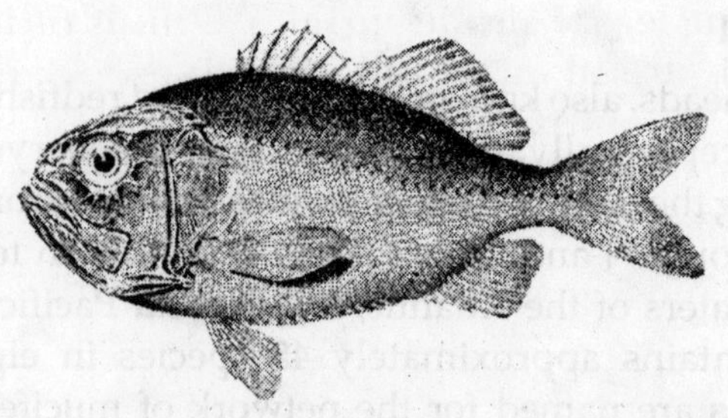

All fins are spinous (excluding the low-slung pectoral fins) and rounded: there is a single dorsal fin with 3-8 spines and 10-19 soft rays; the pelvic fins are thoracic with one spine and 6-7 soft rays; the anal fin has 2-3 spines and 8-12 soft rays; and even the forked caudal fin possesses 4-7 procurrent spines on each lobe. The scales of slimeheads are ctenoid but vary interspecifically; they range from deciduous to adherent.

In most species the ventral scales between the pelvic fin and anus have been modified into a median ridge of large, bony scutes. The lateral line is uninterrupted and fairly obvious; its pores are largely obscured by the scales' well-developed spinules or *ctenii*.

Slimeheads range from a bright brick red with identically shaded fins, to dusky grey or silver, to black with dusky grey to transparent fins. The reds quickly fade to orange following death. Some species (e.g., *Aulotrachichthys latus*) are reported to be bioluminescent, probably via symbiotic bacteria as is found in other beryciform fish. The largest species is the orange roughy at a maximum standard length (SL; a measurement excluding the caudal fin) of 75 cm and a weight of 7 kg; however, most slimeheads are well under 30 cm SL.

Life History

Most slimeheads are sluggish and demersal, spending most of their time near the bottom of continental slopes. Cold, moderate benthopelagic depths (ca. 100-1,500 m) with usually hard, rocky substrates are frequented. The most elongate species are typically the most active and frequent the shallowest depths) for example, the slender roughy (*Optivus elongatus*) is found in photic coastal waters and is associated with rocky reefs. This species is nocturnal and hides in crevices during the day. *Trachichthys australis* is of the same habitus, but is rather deep-bodied and resembles a soldierfish. Both young and adult slimeheads feed primarily upon zooplankton such as mysid shrimp, amphipods, euphausiids, prawns and other crustaceans, as well as larval fish. Slimeheads store energy as extracellular wax esters, which aid the fish in maintaining neutral buoyancy.

Slimehead behaviour is not well studied, but some species sporadically form dense aggregations. In the case of the orange roughy, these aggregations (possibly segregated according to sex) may reach a population density of 2.5/m². The aggregations form in and around geologic structures, such as undersea canyons and seamounts, likely where water movement and mixing is high, ensuring dense concentrations of prey items. The aggregations do not necessarily form for the purpose of spawning; it is thought that the fish cycle through metabolic phases (feeding and resting) and seek areas with ideal hydrologic conditions to congregate during their inactive and active phases.

Observations of orange roughy aggregations during submersible dives have also shown that the fish lose almost all pigmentation while inactive, during which time they are very approachable. The orange roughy's metabolic phases are thought to be related to seasonal variations in the fishes' prey concentrations, with the inactive phase being a means to conserve energy during lean periods.

Slimeheads are non-guarding pelagic spawners; that is, spawning aggregations are formed and the fish release eggs and sperm en masse directly into the water. There is evidence of oceanodromy (seasonal migration) in some species. The fertilized eggs (and later the larvae) are planktonic, floating with the currents until the larvae develop the strength to determine their own way. Only the economically important species have had their reproduction studied in any detail: the larvae and juveniles of Darwin's slimehead are pelagic and frequent rather shallow waters near the coast, whereas in orange roughy the early life stages are apparently confined to deeper water (ca. 200 metres). Slimeheads are very slow-growing and long-lived fish; the orange roughy ranks among the longest-lived animals known, with a maximum reported age of 149 years (however, this age is disputed). Predators of slimeheads are not well known, but include large deep-roving sharks; cutthroat eels; merluccid hakes, and snake mackerels.

Species

There are 45 species in eight genera:

- Genus *Aulotrachichthys*
  - — *Aulotrachichthys latus* (Fowler, 1938).
- Genus *Gephyroberyx*
  - — Darwin's slimehead, *Gephyroberyx darwinii* (Johnson, 1866).
  - — *Gephyroberyx japonicus* (Hilgendorf, 1879).
  - — *Gephyroberyx philippinus* (Fowler, 1938).
- Genus *Hoplostethus*
  - — *Hoplostethus abramovi* (Kotlyar, 1986).
  - — Orange roughy, *Hoplostethus atlanticus* (Collett, 1889).
  - — Black slimehead, *Hoplostethus cadenati* (Quero, 1974).
  - — *Hoplostethus confinis* (Kotlyar, 1980).
  - — *Hoplostethus crassispinus* (Kotlyar, 1980).
  - — *Hoplostethus druzhinini* (Kotlyar, 1986).
  - — *Hoplostethus fedorovi* (Kotlyar, 1986).
  - — *Hoplostethus fragilis* (de Buen, 1959).
  - — *Hoplostethus gigas* (McCulloch, 1914).
  - — Common sawbelly, *Hoplostethus intermedius* (Hector, 1875).
  - — Flintperch, *Hoplostethus japonicus* (Hilgendorf, 1879).
  - — Giant sawbelly, *Hoplostethus latus* (Fowler, 1938).

— *Hoplostethus marisrubri* (Kotlyar, 1986).

— Mediterranean slimehead, *Hoplostethus mediterraneus mediterraneus* Cuvier, *Hoplostethus mediterraneus sonodae* (Kotylar, 1986).

— Silver roughy, *Hoplostethus mediterraneus trunovi* (Kotylar, 1986).

— *Hoplostethus melanopterus* (Fowler, 1938).

— Smallscale slimehead, *Hoplostethus melanopus* (Weber, 1913).

— *Hoplostethus mento* (Garman, 1899).

— *Hoplostethus metallicus* (Fowler, 1938).

— *Hoplostethus mikhailini* (Kotlyar, 1986).

— *Hoplostethus occidentalis* (Woods, 1973).

— *Hoplostethus pacificus* (Garman, 1899).

— *Hoplostethus rifti* (Kotlyar, 1986).

— *Hoplostethus rubellopterus* (Kotlyar, 1980).

— *Hoplostethus shubnikovi* (Kotlyar, 1980).

— *Hoplostethus tenebricus* (Kotlyar, 1980).

— *Hoplostethus vniro* (Kotlyar, 1995).

- Genus *Optivus*

— Slender roughy, *Optivus elongatus* (Gunther, 1859).

- Genus *Paratrachichthys*

— *Paratrachichthys argyrophanus* (Woods, 1961).

— *Paratrachichthys atlanticus* (Collett, 1889).

— *Paratrachichthys fernandezianus* (Gunther, 1887).

— *Paratrachichthys heptalepis* (Gon, 1984).

— *Paratrachichthys novaezelandicus* (Kotlyar, 1980).

— *Paratrachichthys prosthemius* (Jordan & Fowler, 1902).

— *Paratrachichthys pulsator* (Gomon & Kuiter, 1987).

— *Paratrachichthys sajademalensis* (Kotlyar, 1979).

— Sandpaper fish or common roughy, *Paratrachichthys trailli* (Hutton, 1875).

- Genus *Parinoberyx*
  - — *Parinoberyx horridus* (Kotlyar, 1984).
- Genus *Sorosichthys*
  - — Little pineapple fish, *Sorosichthys ananassa* (Whitley, 1945).
- Genus *Trachichthys*
  - — *Trachichthys australis* (Shaw, 1799).

Longfin

The longfins also known as roundheads or spiny basslets are a family, Plesiopidae, of fishes in the order Perciformes. They are elongated fishes, found in the Indian Ocean and western Pacific Ocean.

Species

In some classifications the genus *Notograptus* is split in its own family, Notograptidae, but this article follows FishBase in including it here.

- Genus *Acanthoclinus* (Jenyns, 1841)
  - — Olive rockfish, *Acanthoclinus fuscus* (Jenyns, 1842).
  - — New Zealand rockfish, *Acanthoclinus littoreus* (Forster, 1801).
  - — Stout rockfish, *Acanthoclinus marilynae* (Hardy, 1985).
  - — *Acanthoclinus matti* (Hardy, 1985).
  - — Little rockfish, *Acanthoclinus rua* (Hardy, 1985).
- Genus *Acanthoplesiops* (Regan, 1912)
  - — *Acanthoplesiops echinatus* (Smith-Vaniz & Johnson, 1990).
  - — Hiatt's basslet, *Acanthoplesiops hiatti* (Schultz, 1953).
  - — Scottie, *Acanthoplesiops indicus* (Day, 1888).
  - — *Acanthoplesiops psilogaster* (Hardy, 1985).
- Genus *Assessor* (Whitley, 1935)
  - — Yellow devilfish, *Assessor flavissimus* (Allen & Kuiter, 1976).
  - — Blue devilfish, *Assessor macneilli* (Whitley, 1935).

— *Assessor randalli* (Allen & Kuiter, 1976).

- Genus *Beliops* (Hardy, 1985)
  — *Beliops batanensis* (Smith-Vaniz & Johnson, 1990).
  — *Beliops xanthokrossos* (Hardy, 1985).
- Genus *Belonepterygion* (McCulloch, 1915)
  — *Belonepterygion fasciolatum* (Ogilby, 1889).
- Genus *Calloplesiops* (Fowler and Bean, 1930)
  — Comet, *Calloplesiops altivelis* (Steindachner, 1903).
  — *Calloplesiops argus* (Fowler & Bean, 1930).
- Genus *Fraudella* (Whitley, 1935)
  — *Fraudella carassiops* (Whitley, 1935).
- Genus *Notograptus*
  — Shark Bay eel-blenny, *Notograptus gregoryi* (Whitley, 1941).
  — Spotted eel blenny, *Notograptus guttatus* (Gunther, 1867).
  — *Notograptus kauffmani* (Tyler & Smith, 1970).
  — *Notograptus livingstonei* (Whitley, 1931).
- Genus *Paraplesiops* (Bleeker, 1875)
  — *Paraplesiops alisonae* (Hoese & Kuiter, 1984).
  — Eastern blue devil, *Paraplesiops bleekeri* (Gunther, 1861).
  — Blue devil, *Paraplesiops meleagris* (Peters, 1869).
  — *Paraplesiops poweri* (Ogilby, 1908).
  — *Paraplesiops sinclairi* (Hutchins, 1987).
- Genus *Plesiops* (Oken, 1817)
  — Bluegill longfin, *Plesiops corallicola* (Bleeker, 1853).
  — *Plesiops facicavus* (Mooi, 1995).
  — *Plesiops genaricus* (Mooi & Randall, 1991).
  — Threadfin longfin, *Plesiops gracilis* (Mooi & Randall, 1991).
  — *Plesiops insularis* (Mooi & Randall, 1991).
  — *Plesiops malalaxus* (Mooi, 1995).

— Spotted longfin, *Plesiops multisquamata* (Inger, 1955).
— Moustache longfin, *Plesiops mystaxus* (Mooi, 1995).
— *Plesiops nakaharai* (Tanaka, 1917).
— *Plesiops auritus* (Mooi, 1995).
— *Plesiops cephalotaenia* (Inger, 1955).
— Crimsontip longfin, *Plesiops coeruleolineatus* (Ruppell, 1835).
— Whitespotted longfin, *Plesiops nigricans* (Ruppell, 1828).
— Sharp-nosed longfin, *Plesiops oxycephalus* (Bleeker, 1855).
— *Plesiops polydactylus* (Mooi, 1995).
— *Plesiops thysanopterus* (Mooi, 1995).
— *Plesiops verecundus* (Mooi, 1995).

- Genus *Steeneichthys* (Allen and Randall, 1985)
  — *Steeneichthys nativitatus* (Allen, 1997).
  — *Steeneichthys plesiopsus* (Allen & Randall, 1985).
- Genus *Trachinops* (Gunther, 1861)
  — *Trachinops brauni* (Allen, 1977).
  — *Trachinops caudimaculatus* (McCoy, 1890).
  — *Trachinops noarlungae* (Glover, 1974).
  — Eastern hulafish, *Trachinops taeniatus* (Gunther, 1861).

Round Whitefish

Round Whitefish (*Prosopium cylindraceum*) is a freshwater species of fish that is found in lakes from Alaska to New England, including the Great Lakes. It is has an olive-brown back with light silvery sides and underside and its size is generally between 9 and 19 inches long. They are bottom feeders, feeding mostly on invertebrates, such as crustaceans, insect larvae, and fish eggs. Other common names of the round whitefish are menominee, pilot fish, frost fish, round-fish, and menominee whitefish. The common name "round whitefish" is also sometimes used to describe *Coregonus huntsmani*, a salmonid more commonly known as the Atlantic whitefish.

While it was once common, numbers have been decreasing in the last century due to a number of possible causes. The round whitefish is now protected in some states, such as New York, under the Endangered Species Act from harvest or possession.

Rudd

The Rudd (*Scardinius erythrophthalmus*) is a small fish, a widespread member of the family Cyprinidae.

The rudd is a bentho-pelagic freshwater fish, widely spread in Europe and middle Asia, around the basins of the North, Baltic Black, Caspian and Aral seas. It has been artificially introduced to Ireland, USA, Morocco, Madagascar, Tunisia, New Zealand, Canada and Spain.

Morphologically, this species is very similar to the Roach (*Rutilus rutilus*), with which it can be easily confused. It can be identified by eye colour (deep blood-red as opposed to yellow) or counting the soft rays in the dorsal fin (8-9 compared to 10-12). Confusingly, these species can hybridise, producing intermediate forms. The rudd can also hybridise with the carp bream *Abramis brama*.

In New Zealand and Canada it is considered a pest fish due to impacts on native species.

Rudderfish

The rudderfish, *Centrolophus niger*, is a medusafish, the only member of the genus *Centrolophus* found in all tropical and temperate oceans of the world, at depths of from 50 to 1,000 m. Its length is from 60 to 150 cm.

The rudderfish is a moderately elongate blunt-headed fish with long low dorsal and anal fins and small pectoral and pelvic fins. The body is covered in small, soft, easily-shed scales.

Its colour is a uniform dusky brown or black, and the fish is a midwater-living species.

Ruffe

The Ruffe (*Gymnocephalus cernuus*) is a freshwater fish found in temperate regions of Europe and northern Asia. It has been introduced into the Great Lakes of North America, reportedly with unfortunate results.

The ruffe is about 100mm (10cm) in total length. It lives in the deeper waters of lakes and quiet pools, or at the margins of streams, preferring a sandy or gravel bottom. It eats mainly invertebrates from the bottom of the water, though it will take small fish and some plant matter. It is in turn preyed on by larger fish.

Sabertooth Fish

Sabertooth fishes (also spelt sabretooth) are small, fierce-looking deep-sea aulopiform fish comprising the family Evermannellidae. The family is small, with just eight species in three genera represented; they are distributed throughout tropical to subtropical waters of the Atlantic, Indian, and Pacific Ocean.

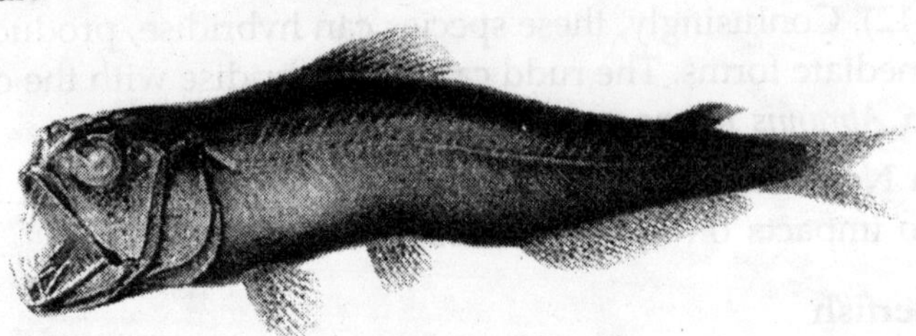

These fishes are appropriately named for their oversized, recurved palatine teeth redolent of the saber-toothed cats. The family name "Evermannellidae" was given in honour of Dr. Barton Warren Evermann, a noted ichthyologist, naturalist and director of the California Academy of Sciences.

Physical Description

Sabertooth fishes have moderately elongate and compressed bodies which lack normal scales. The head is large and blunt; the terminal mouth is large, and it is lined with slender palatine teeth, the frontmost of which are greatly enlarged and curve inward slightly. A number of shorter, straighter teeth accompany these fang-like teeth. The tongue is toothless. The eyes range in size from small to large; they are tubular in structure and point upwards. The lateral line runs uninterrupted. The vertebrae number 44-54 and there are three discrete bands of muscle tissue (epaxial, midlateral, and hypaxial) present in the caudal region. The swim bladder is absent and the stomach is highly distensible.

There is a single high dorsal fin (with 10-13 rays) originating slightly before the thoracic pelvic fins. The anal fin (26-37 rays) is the largest of the fins, and runs along the posterior half of the fish, tapering in height towards the emarginate caudal fin. A small adipose fin is also present.

The pectoral fins (11-13 rays) are positioned rather low on the body. All fins are spineless and lightly pigmented in shades of brown.

Sabertooth fish are usually a drab, light to dark brown when preserved; however, a brassy green iridescence is seen on the flanks, cheeks, and ocular region of well-preserved specimens. The naked skin is easily torn. The Atlantic sabertooth (*Coccorella atlantica*) is the largest species, at up to 18.5 centimetres standard length.

Life History

Almost nothing is known of the biology and ecology of evermannellids. They are active, visual predators and confine themselves to the mesopelagic zone, c. 200-1,000 metres down; they are most commonly trawled from between 200-400 metres. At these depths there is extremely little to no light; the view from below is like the sky at twilight. The sabertooh fish use their telescopic, upward-pointing eyes—which are thus adapted for improved terminal vision at the expense of lateral vision—to pick out squid, cuttlefish, and smaller fish that are silhouetted against the gloom above them.

Their distensible stomachs allow sabertooth fish to swallow prey larger than themselves; their recurved teeth likely function in a manner similar to a snake's, preventing a captured fish from backing out and helping to guide the fish down the sabertooth's pharynx. Sabertooth fish are solitary animals; it is not known whether they undergo diel vertical migrations.

Their reproductive habits are poorly studied; they are assumed to be non-guarding, pelagic spawners. True synchronous hermaphroditism with external fertilization is known in *Evermannella indica* and *Odontostomops normalops,* and the former species appears to spawn throughout the year. Sabertooth fish larvae are planktonic and have long snouts and oblong eyes before metamorphosis. Both larvae and juveniles remain at shallower depths of 50-100 metres, descending to deeper water with age.

Species

There are eight species in three genera:

- Genus *Coccorella*
  - — Atlantic sabretooth, *Coccorella atlantica* (Borodin, 1931).
  - — *Coccorella atrata* (Alcock, 1894).
- Genus *Evermannella*
  - — *Evermannella ahlstromi* (Johnson & Glodek, 1975).
  - — Balbo sabretooth, *Evermannella balbo* (Risso, 1820).
  - — *Evermannella indica* (Brauer, 1906).
  - — *Evermannella megalops* (Johnson & Glodek, 1975).
  - — Indian sabertooth, *Evermannella melanoderma* (Parr, 1928).
- Genus *Odontostomops*
  - — Undistinguished sabretooth, *Odontostomops normalops* (Parr, 1928).

Sailfish

Sailfishes (genus *Istiophorus*) are fish living in all the oceans of the world. They are blue to grey in colour and have a characteristic sail (dorsal fin) on top, which often stretches the entire length of the back. Another notable characteristic is the elongated bill, resembling that of a swordfish.

All sailfish species grow quickly, reaching 1.2 to 1.5 metres in length in a single year, and feed on the surface or at mid-depths on smaller pelagic fishes and squid. Individuals have been clocked at speeds of up to 68.5 mph, making them the fastest fish in the ocean. Generally, sailfish do not grow to more than 10 feet in length and rarely weigh over 200 pounds, although larger specimens have been seen off the shores of Costa Rica.

The sail is normally kept folded down and to the side when swimming, but it may be raised when the sailfish feels threatened or excited, making it appear much larger than it actually is. This tactic has also been observed during feeding, when a group of sailfish use their sails to "herd" a school of fish or squid.

Species

- Atlantic sailfish, *Istiophorus albicans*.
- Indo-Pacific sailfish, *Istiophorus platypterus*.

Salamanderfish

The salamanderfish, *Lepidogalaxias salamandroides*, is the only species in the family Lepidogalaxiidae. First described in 1961, it is only found in acidic pools of water in the heathland peat flats of southwest Western Australia. The species is tiny, measuring approximately 7 cm in length. It is also unusual for its ability to survive desiccation by burrowing into sand when the pools it lives in periodically evaporate.

Salmon Shark

The salmon shark, *Lamna ditropis*, is a shark species occurring in the North Pacific Ocean. As an apex predator, the salmon shark feeds on salmon, and also on squid, sablefish, and herring.

Salmon shark are remarkable for their ability to maintain body temperature, known as homeothermy, and an as-yet unexplained variability in the sex ratio between the eastern North Pacific and Western North Pacific.

Biology

Adult salmon sharks are dark grey to black over most of the body, with a white underside with darker blotches. Juveniles are similar in appearance but generally lack blotches. The snout is short and cone-shaped, and the overall appearance is similar to a small great white shark.

Salmon shark generally grow to between 200 and 260 cm in length and weigh up to 220 kg. Males appear to reach a maximum size which is slightly smaller than females. Unconfirmed reports exist of salmon shark reaching as much as 4.3 m however the largest confirmed reports indicate a maximum total length of approximately 3 m.

Reproduction

Salmon shark are ovoviviparous, with a litter size of 2 to 5 pups. As with other Lamniformes shark species, salmon sharks are oophagous, with embryos feeding on the ova produced by the mother.

Females reach sexual maturity at 8 to 10 years, while males general mature by age 5. Reproduction timing is not well understood, however it is believed to be on a two years cycle with mating occurring in the late summer or early fall. Gestation is approximately 9 months. Some reports indicate that the sex ratio at birth may be 2.2 males per female, however the prevalence of this is not known.

Homeothermy

As with only a few other species of fish, salmon shark have the ability to regulate their body temperature. This is accomplished by vascular countercurrent heat exchangers, known as *rete mirabilia*, Latin for "wonderful nets." Blood moving towards extremities flows near colder returning blood, resulting in heat transfer. The returning blood is warmed, keeping the core of the animal heated. This ability is believed to help the salmon shark exploit prey in a wider range of water temperatures.

Range and Distribution

The salmon shark is a coastal-littoral and epipelagic shark that prefers boreal to cool temperate waters. It is common in continental offshore waters but range inshore to just off beaches. They occur singly or in schools or feeding aggregations of several individuals.

Salmon shark occur in the northern Pacific Ocean, in both coastal waters and in the open ocean. Animals are believed to range as far south as the Sea of Japan and southern California, and as far north as 65 degrees north in Alaska. Individuals have been observing diving as deeply as 668 m , however they are believed to spend most of their time in epipelagic waters.

Regional Differences

Age and sex composition differences have been observed between populations in the eastern and western North Pacific. Eastern populations are dominated by females, while the western populations are predominately male. It is not known if these distinctions stem from genetically distinct stocks, or if the segregation occurs as part of salmon shark growth and development.

Human Interactions

There is no current commercial fishery for salmon shark, however, they are occasionally caught as bycatch in commercial salmon gillnet fisheries where they are usually discarded. Commercial fisheries regard salmon sharks as nuisances, since they can damage fishing gear, and consume portions of the commercial catch. There are some reports of fishermen deliberately injuring salmon sharks.

Sport fishermen fish for salmon sharks in Alaska. Alaskan fishing regulations limit the catch of salmon shark to two sharks per person per year.

The flesh of the fish is utilised for human consumption, and in the Japanese city of Kesennuma, Miyagi, the heart is considered a delicacy for use in sashimi.

Although salmon sharks are thought to be capable of

injuring humans, there are few if any attacks on humans. There are, however, reports of divers encountering salmon sharks as well as salmon sharks bumping fishing vessels. These reports, however, may need positive identification of the shark species involved.

Sandbar Shark

The sandbar shark, *Carcharhinus plumbeus*, comes from the Carcharhinidae family of sharks, also called requiem sharks.

The sandbar shark is also called the thickskin shark or brown shark. It is one of the biggest coastal sharks in the world, and is closely related to the dusky shark, the bignose shark, and the bull shark.

Its dorsal fin is triangular and very high, and weighs as much as 18 per cent of the shark's whole body. Sandbar sharks usually have heavy-set bodies and rounded snouts that are shorter than the average shark's snout. Their upper teeth have broadly uneven cusps with sharp edges. Its second dorsal fin and anal fin are close to the same height. Females can grow to 7 or 8 feet, males up to 6 feet. Its body colour can vary from a bluish to a brownish grey to a bronze, with a white or pale underside. Sandbar sharks swim alone or gather in sex-segregated schools that vary in size. They are most active at night, at dawn, and at dusk.

The sandbar shark, true to its nickname, is commonly found over muddy or sandy bottoms in shallow coastal waters such as bays, estuaries, harbours, or the mouths of rivers, but it also swims in deeper waters (200 m or more) as well as intertidal zones. Sandbar sharks are found in tropical to temperate waters worldwide; in the western Atlantic they range from Massachusetts to Brazil. Juveniles are common to abundant in the lower Chesapeake Bay, which is probably one of the most important nursery grounds on the United States east coast for this species.

The sandbar shark's main predator is man. Natural predators are the tiger sharks, and rarely by great white sharks.

The sandbar sharks, and other requiem sharks, prey on finfish rays, bottom dwelling animals, seabirds and turtles.

Sandbar sharks are viviparous. The embryos are supported in placental yolk sac inside the mother. The female reproduces every two years. They give birth to 8 to 10 young. They carry the young for 1 year before birth.

Sandburrower

The sandburrowers or simply burrowers are a family, Creediidae, of fishes in the order Perciformes.

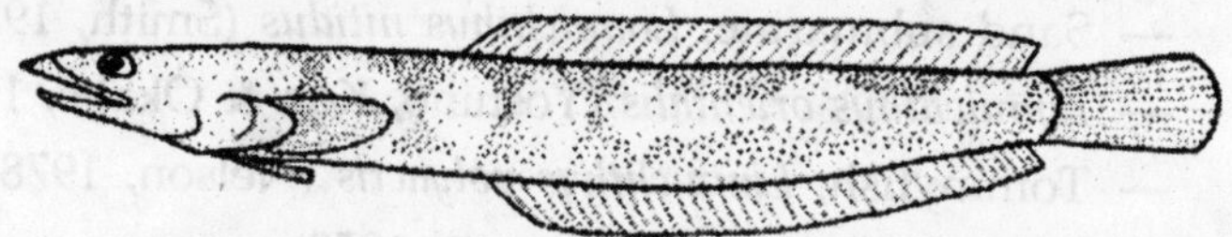

They are native to coastal waters the Indian and Pacific Oceans. They are very small fishes: with the exception of the larger Donaldson's sandburrower, *Limnichthys donaldsoni*, most species reach only 3 to 7 cm in length. They live in shallow waters close to the shore, burrowing into sandy areas swept by currents or by surf.

Species

There are seventeen species in seven genera:

- Genus *Apodocreedia* (de Beaufort, 1948)
  - — Longfin burrower, *Apodocreedia vanderhorsti* (de Beaufort, 1948).
- Genus *Chalixodytes* (Schultz, 1943)
  - — Sand dart, *Chalixodytes chameleontoculis* (Smith, 1957).
  - — Saddled sandburrower, *Chalixodytes tauensis* (Schultz, 1943).
- Genus *Creedia* (Ogilby, 1898)
  - — *Creedia alleni* (Nelson, 1983).
  - — *Creedia bilineatus* (Shimada & Yoshino, 1987).
  - — Slender sand-diver, *Creedia haswelli* (Ramsay, 1881).

— Half-scaled sand-diver, *Creedia partimsquamigera* (Nelson, 1983).

- Genus *Crystallodytes* (Fowler, 1923)
  - — South Pacific Sandburrower, *Crystallodytes cookei* (Fowler, 1923).
  - — *Crystallodytes pauciradiatus* (Nelson & Randall, 1985).
- Genus *Limnichthys* (Waite, 1904)
  - — Donaldson's sandburrower, *Limnichthys donaldsoni* (Schultz, 1960).
  - — Barred sand burrower, *Limnichthys fasciatus* (Waite, 1904).
  - — Sand submarine, *Limnichthys nitidus* (Smith, 1958).
  - — *Limnichthys orientalis* (Yoshino, Kon & Okabe, 1999).
  - — Tommyfish, *Limnichthys polyactis* (Nelson, 1978).
  - — *Limnichthys rendahli* (Parrott, 1958).
- Genus *Schizochirus* (Waite, 1904)
  - — *Schizochirus insolens* (Waite, 1904).
- Genus *Tewara* (Griffin, 1933)
  - — New Zealand sand diver, *Tewara cranwellae* (Griffin, 1933).

Sanddab

A sanddab is an edible flatfish of the genus *Citharichthys*. Like other flatfish, they have both eyes on the same side of the head, most often on the left side. There are multiple species of sanddab, though the commonest by far is the Pacific sanddab, *Citharichthys sordidus*. They are a dull light-brown, and are mottled with brown or black; sometimes yellow or orange.

Sand eel

Sand eel or sandeel is the common name used for a considerable number of species of fish. Most of them are sea fish of the genera *Hyperoplus* (greater sandeels), *Gymnammodytes* or *Ammodytes*. Many species are found off the western coasts of Europe from Spain to Scotland, and in the Mediterranean and Baltic Seas.

The three genera listed above all fall within the family Ammodytidae, the sandlances. Members of these genera found in other oceans are not usually called Sand Eels, and species from other parts of the world that are known as sandeels are usually less closely related. None of the Sand Eels are related to the true eels.

Sandeels are an important food source for diving birds, including puffins and shags.

Sandeels burrow in the sand to escape from predators.

Commercial Fishing

Traditionally they have been little exploited for human food, but are a major target of "industrial fishing" for animal feed and fertilizer, particularly in the North Sea.

Increasing fishing of them is thought to be causing problems for some of their natural predators, especially the auks which take them in deeper water.

Sand Lance

A sand lance or sandlance is a fish belonging to the family Ammodytidae. Several species of sand lance are commonly known as "sand eels" or "sandeels", though they are not related to true eels. Another variant name is launce, and all names of the fish are references to its slender body and pointed snout. The family name (and genus name, *Ammodytes*) means "sand burrower", which describes the sand lance's habit of burrowing into sand to avoid tidal currents.

Sand lances are most commonly encountered by fishermen in the North Pacific and North Atlantic, but are found in oceans throughout the world. These fish do not have pelvic fins and do not develop a swim bladder, staying true to their bottom-dwelling habit as adults. Both adult and larval sea lances primarily feed on copepods. Larval forms of this fish are perhaps the most abundant of all fish larvae in areas such as the northwest Atlantic, serving as a major food item for cod, salmon, and other commercially important species.

As adults, sand lances are harvested commercially in some areas (primarily in Europe), leading to direct human competition with diving birds such as puffins, auks, and cormorants.

Some species are inshore coastal dwellers, and digging for sand lances to use as a bait fish has been a popular pastime in coastal areas of Europe and North America. Other species are deep-water dwellers, some of which have only recently been described to science, and most of which lack common names.

Species

There are twenty-three species in seven genera:

- Genus *Ammodytes*
  - — American sand lance, *Ammodytes americanus* (DeKay, 1842).
  - — Northern sand lance, *Ammodytes dubius* (Reinhardt, 1837).
  - — Pacific sand lance, *Ammodytes hexapterus* (Pallas, 1814).
  - — Lesser sand-eel, *Ammodytes marinus* (Raitt, 1934).
  - — Pacific sandeel, *Ammodytes personatus* (Girard, 1856).
  - — Small sandeel, *Ammodytes tobianus* (Linnaeus, 1758).
- Genus *Ammodytoides*
  - — Gill's sand lance, *Ammodytoides gilli* (Bean, 1895).
  - — *Ammodytoides kimurai* (Ida & Randall, 1993).
  - — Pitcairn Sandlance, *Ammodytoides leptus* (Collette & Randall, 2000).
  - — *Ammodytoides pylei* (Randall, Ida & Earle, 1994).
  - — Scaly sandlance, *Ammodytoides renniei* (Smith, 1957).
  - — *Ammodytoides vagus* (McCulloch & Waite, 1916).
- Genus *Bleekeria*
  - — *Bleekeria kallolepis* (Gunther, 1862).
  - — *Bleekeria mitsukurii* (Jordan & Evermann, 1902).
  - — *Bleekeria viridianguilla* (Fowler, 1931).

- Genus *Gymnammodytes*
  - — Cape sandlance, *Gymnammodytes capensis* (Barnard, 1927).
  - — Mediterranean sand eel, *Gymnammodytes cicerelus* (Rafinesque, 1810).
  - — Smooth sandeel, *Gymnammodytes semisquamatus* (Jourdain, 1879).
- Genus *Hyperoplus*
  - — Greater sand-eel, *Hyperoplus immaculatus* (Corbin, 1950).
  - — Great sandeel, *Hyperoplus lanceolatus* (Le Sauvage, 1824).
- Genus *Lepidammodytes*
  - — *Lepidammodytes macrophthalmus* (Ida, Sirimontaporn & Monkolprasit, 1994).
- Genus *Protammodytes*
  - — *Protammodytes brachistos* (Ida, Sirimontaporn & Monkolprasit, 1994).
  - — *Protammodytes sarisa* (Robins & Bohlke, 1970).

Sandperch

The sandperches are a family, Pinguipedidae, of fishes in the order Perciformes.

Species

There are 63 species in seven genera:

- Genus *Kochichthys* (Kamohara, 1961).
  - — *Kochichthys flavofasciata* (Kamohara, 1936).
- Genus *Parapercis* (Bleeker, 1863)
  - — Whitespot sandsmelt, *Parapercis alboguttata* (Gunther, 1872).
  - — Barred grubfish, *Parapercis allporti* (Gunther, 1876).
  - — *Parapercis atlantica* (Vaillant, 1887).
  - — *Parapercis aurantiaca* (Doderlein, 1884).

— *Parapercis australis* (Randall, 2003).
— Redbanded weever, *Parapercis binivirgata* (Waite, 1904).
— *Parapercis biordinis* (Allen, 1976).
— *Parapercis cephalopunctata* (Seale, 1901).
— Latticed sandperch, *Parapercis clathrata* (Ogilby, 1910).
— *Parapercis colemani* (Randall & Francis, 1993).
— Blue cod, *Parapercis colias* (Forster, 1801).
— Cylindrical sandperch, *Parapercis cylindrica* (Bloch, 1792).
— *Parapercis decemfasciata* (Franz, 1910).
— Doublespot grubfish, *Parapercis diplospilus* (Gomon, 1980).
— *Parapercis dockinsi* (McCosker, 1971).
— *Parapercis elongata* (Fourmanoir, 1967).
— Threadfin sandperch, *Parapercis filamentosa* (Steindachner, 1878).
— *Parapercis flavescens* (Fourmanoir & Rivaton, 1979).
— *Parapercis flavolabiata* (Johnson, 2006).
— *Parapercis fuscolineata* (Fourmanoir, 1985).
— Yellow weaver, *Parapercis gilliesii* (Hutton, 1879).
— Wavy grubfish, *Parapercis haackei* (Steindachner, 1884).
— Speckled sandperch, *Parapercis hexophtalma* (Cuvier, 1829).
— *Parapercis kamoharai* (Schultz, 1966).
— Y-Barred Sandperch, *Parapercis lata* (Randall & McCosker, 2002).
— Narrow barred grubfish, *Parapercis macrophthalma* (Pietschmann, 1911).
— Harlequin sandperch, *Parapercis maculata* (Bloch & Schneider, 1801).
— *Parapercis maritzi* (Anderson, 1992).
— Black dotted sand perch, *Parapercis millepunctata* (Gunther, 1860).

— *Parapercis mimaseana* (Kamohara, 1937).
— Gold-birdled sandsmelt, *Parapercis multifasciata* (Doderlein, 1884).
— Redbarred sandperch, *Parapercis multiplicata* (Randall, 1984).
— *Parapercis muronis* (Tanaka, 1918).
— Barred sandperch, *Parapercis nebulosa* (Quoy & Gaimard, 1825).
— *Parapercis okamurai* (Kamohara, 1960).
— *Parapercis ommatura* (Jordan & Snyder, 1902).
— Harlequin sandsmelt, *Parapercis pulchella* (Temminck & Schlegel, 1843).
— *Parapercis punctata* (Cuvier, 1829).
— Spotted sandperch, *Parapercis punctulata* (Cuvier, 1829).
— *Parapercis quadrispinosus* (Weber, 1913).
— Spotted grubfish, *Parapercis ramsayi* (Steindachner, 1883).
— Smallscale grubfish, *Parapercis robinsoni* (Fowler, 1929).
— *Parapercis roseoviridis* (Gilbert, 1905).
— *Parapercis rufa* (Randall, 2001).
— Redspotted sandperch, *Parapercis schauinslandii* (Steindachner, 1900).
— Grub fish, *Parapercis sexfasciata* (Temminck & Schlegel, 1843).
— *Parapercis sexlorata* (Johnson, 2006).
— Blackflag sandperch, *Parapercis signata* (Randall, 1984).
— *Parapercis simulata* (Schultz, 1968).
— U-mark sandperch, *Parapercis snyderi* (Jordan & Starks, 1905).
— Somali sandperch, *Parapercis somaliensis* (Schultz, 1968).
— White-streaked grubfish, *Parapercis stricticeps* (De Vis, 1884).

— *Parapercis striolata* (Weber, 1913).
— Reticulated sandperch, *Parapercis tetracantha* (Lacepede, 1802).
— Yellowbar sandperch, *Parapercis xanthozona* (Bleeker, 1849).

• Genus *Pinguipes* (Cuvier in Cuvier and Valenciennes, 1829).
— *Pinguipes brasilianus* (Cuvier, 1829).
— *Pinguipes chilensis* (Norman, 1937).

• Genus *Prolatilus* (Gill, 1865)
— Tilefish, *Prolatilus jugularis* (Valenciennes, 1833).

• Genus *Pseudopercis* (Miranda Ribeiro, 1903)
— Namorado sandperch, *Pseudopercis numida* (Miranda-Ribeiro, 1903).
— *Pseudopercis semifasciata* (Cuvier, 1829).

• Genus *Ryukyupercis* (Imamura & Yoshino, 2007).
— Rosy grubfish, *Ryukyupercis gushikeni* (Yoshino, 1975).

• Genus *Simipercis* (Johnson & Randall 2006).
— *Simipercis trispinosa* (Johnson & Randall 2006).

Sand Stargazer

Sand stargazers are blennioids; perciform marine fish of the odd family Dactyloscopidae. Found in temperate to tropical waters of North and South America; some may also inhabit brackish environments. There are 44 species in nine genera represented, the giant sand stargazer (*Dactylagnus mundus*) being the largest at 15 centimetres in length; all other species are under 10 centimetres.

These blennies are named well: sand stargazers have protruding eyes on the top of their heads, fixed in an upward gaze, and may be on stalks. Their large mouths are also upturned. The dorsal fin is long and may or may not be continuous, with 7-23 spines; the pelvic fins are situated below the throat and possess one spine. The anal fin is equally long and flowing. The mouth is fringed, and like the upper edge of the operculum (the gill cover), this fringe is divided into finger-like structures. The body is greatly elongate and colouration is generally drab.

As their name would suggest, sand stargazers spend most of their time buried in sandy substrates waiting for unsuspecting prey; only the eyes, nose and mouth are usually visible.

Their mode of respiration is also unique among the blennioids, utilising a *branchiostegal* rather than *opercular* pump; this is thought to be an adaptation to their largely sedentary, obscured lives. Sand stargazers generally stay within shallow (< 10 metres) intertidal zones in areas protected from surges. Small invertebrates and fishes make up the bulk of the sand stargazer's diet.

The family name *Dactyloscopidae* derives from the Greek words *daktylos* meaning "finger" (a reference to the divided mouth and operculum fringes) and *skopein* meaning "to watch".

Sarcastic Fringehead

The sarcastic fringehead, *Neoclinus blanchardi*, is a fish of the suborder Blenioidea. They can be up to thirty cm in length and are the largest fish in the suborder. Their geographic range is from San Francisco, California to central Baja California, Mexico, and their depth range is from 3 to 73 m. Blennoids are mostly scaleless fish with large pectoral fins and reduced pelvic fins. With highly compressed bodies, some may be so elongate as to appear eel-like. They prefer to hide inside various objects or crevices. After the female spawns under a rock or in clam burrows the male guards the eggs.

Sargassumfish

The sargassumfish, *Histrio histrio* (literally, "The Actor"), is a frogfish of the family Antennariidae, found close to the surface amongst floating sargassum weed in all subtropical oceans. Its length is up to 20 cm.

The sargassumfish has numerous fleshy weed-like dermal

appendages which allow it to blend in with the floating sargassum weed in which it is usually found. It is a deep-bodied shapeless species, with long and thickened pectoral and pelvic fins which serve as limbs allowing the fish to 'walk' amongst the seaweed fronds.

They lie in wait for passing crustaceans and small fish, lunging out suddenly to snap up their next meal.

Sauger

The Sauger (*Sander canadensis*) is a freshwater Perciform fish of the family Percidae which resembles its close relative the walleye. Saugers, however, are usually smaller and will tolerate waters of higher turbidity than will the walleye. In many parts of their range, saugers are sympatric with walleyes. They may be distinguished from walleyes by the distinctly spotted dorsal fin, by the lack of a white splotch on the anal fin, and by their generally more brassy colour. The average sauger in an angler's creel is 300 to 400 g (0.75 to 1 lbs) in weight but the world record was 8.1 kg (17 lbs, 12 ounces.) Saugers are more typical of rivers whereas walleyes are more common in lakes and reservoirs. The sauger is highly prized as a food fish.

Hybridisation between saugers and walleyes is not unknown; the hybrids, referred to as saugeyes, exhibit traits of both species. Being intermediate in appearance between the two species, saugeyes are sometimes difficult to differentiate, but they generally carry the dark blotches characteristic of the sauger.

Saury

Sauries are fish of the family Scomberesocidae. There are two genera, each containing two species.

Sauries are marine epipelagic fish which live in tropical and temperate waters. These fish often jump while swimming near the surface, skimming the water. The jaws of sauries are beak-like, ranging from long, slender beaks to relatively short ones with lower jaw only slightly elongated.

The mouth openings of sauries, however, are relatively small and the jaws are weakly toothed. A row of small finlets behind the dorsal and anal fins is also a feature of sauries. An unusual feature of these fish is that they lack swim bladders. Sauries grow to a maximum length of about 46 cm. They are harvested commercially as a food fish.

Sauries first appear in the fossil record in the upper Tertiary, Miocene. The name Scomberesocidae is derived from the Greek, skombros = tunny/mackerel, and esox = nursery of salmon. Pacific saury are consumed often in Japanese and Korean cuisine. The fish is usually grilled.

Species

- Genus *Cololabis*
  - — *Cololabis adocetus*
  - — *Cololabis saira* (Pacific saury)
- Genus *Scomberesox* (sauries and skippers)
  - — *Scomberesox forsteri* (skipper garfish)
  - — *Scomberesox saurus* (Atlantic saury)

Sawfish

Sawfishes are a family of marine animals related to sharks and rays. Their most striking appearance is a long, toothy snout. They possess a cartilaginous skeleton and no swim bladder. They are the sole family Pristidae of the order Pristiformes, from the Greek and Latin *pristis* meaning "sawfish".

They are not to be confused with sawsharks (order Pristiophoriformes), which have a similar physical appearance.

All species of sawfish are considered endangered, or critically endangered and international trade is banned.

Description

The most eye-catching feature of the sawfish is their saw-like snout, called a rostrum. The rostrum is covered with motion- and electro-sensitive pores that allow sawfish to detect movement and even heartbeats of buried prey in the ocean floor. The rostrum acts like a metal detector as the sawfish hovers over the bottom, looking for hidden food. It is also used as a digging tool to unearth buried crustaceans. When a suitable prey swims by, the normally lethargic sawfish will spring from the bottom and slash at it furiously with its saw.

This generally stuns or injures the prey sufficiently for the sawfish to devour it without much resistance. Sawfish have also been known to defend themselves with their rostrum, against predators (like sharks) and intruding divers. The "teeth" protruding from the rostrum are not real teeth, but modified denticle scales (The scales of a sawfish have a similar structure to its teeth, confusing the distinction somewhat).

The body and head of a sawfish is flat as they spend most of their time lying on the sea floor. Like rays, the sawfish's mouth and nares are located on its flat underside. The mouth is lined with small, dome-shaped teeth for eating small fish and crustaceans; though sometimes the fish swallows them whole. Sawfishes breathe with two spiracles just behind the eyes that draw water to the gills. The skin is covered with tiny dermal denticles (skin-teeth) that gives the fish a rough texture. Sawfishes are usually light grey or brown; the smalltooth sawfish, *Pristis pectinata*, appears olive green.

Like other elasmobranchs, sawfishes lack a swim bladder and use a large, oil-filled liver instead to keep them buoyant. Their skeleton is made of cartilage.

The eyes of sawfish are undeveloped due to their muddy habitat. The rostrum is the main sensory device.

The intestines are shaped like a corkscrew, called a spiral-valve.

The smallest sawfish is the 1.4 m (4.6 foot) dwarf sawfish, *Pristis clavata*, a species much smaller than most other sawfish. The largest species seem to be the largetooth sawfish, *Pristis microdon* and the southern sawfish, *Pristis perotteti*, both of which can exceed 7 m (23 feet) in length. One southern sawfish, whose length for some reason went unmeasured, was said to have weighed 2,455 kg (5,400 lb).

Distribution and Habitat

Sawfishs are found in tropical and subtropical areas around Africa and Australia and in the Caribbean, and frequently ascend far into rivers. They are also found in bays and estuaries.

Sawfishs live only in shallow, muddy water and can be found in both freshwater and saltwater. Most prefer river mouths and freshwater systems. All sawfish have the ability to traverse between fresh and saltwater, and often do so.

Behaviour

Sawfishes are nocturnal, usually sleeping during the day, hunting at night. Despite fearsome appearances, they are gentle fishes and will not attack humans unless provoked or surprised. The smalltooth sawfish is well known by fishermen as a prize game fish because of the fight it puts up once hooked. Capturing sawfishes is illegal in the United States and Australia.

Reproduction

Little is known about the reproduction habits of the sawfish. Each individual lives around 25 to 30 years, and matures at 10 years.

Females give live birth to pups, whose semi-hardened rostrum is covered with a rubbery envelope. This prevents the pup from injuring its mother during birth. The rubbery envelope eventually disintegrates and falls off.

The sawfish is estimated to mate once every two years, with an average litter of around eight pups.

Species

- Genus *Anoxypristis*
  - — Knifetooth sawfish, *Anoxypristis cuspidata* (Latham, 1794).

Also known as the narrow or pointed sawfish. Lives in muddy areas, appears grey.

- Genus *Pristis*
  - — Dwarf sawfish, *Pristis clavata* (Garman, 1906).

Also known as the Queensland sawfish. Inhabits muddy bays and estuaries along the northern coast of Australia. Relatively small compared to other species, only around 1.4 m.

— Largetooth sawfish, *Pristis microdon* (Latham, 1794).

Inhabits freshwater systems and have been found deep inland. The largest Australian freshwater fish.

— Smalltooth sawfish, *Pristis pectinata* (Latham, 1794).

Also known as the wide sawfish. Lives in muddy areas, appears green or bluish-grey. Found Also lives in the Caribbean and around the Australian and African coastlines.

— Large-tooth sawfish, *Pristis perotteti* (Muller & Henle, 1841).

Lives around the Caribbean and Central American coastline. Seriously endangered in Central and South America, especially Lake Nicaragua.

— Common sawfish, *Pristis pristis* (Linnaeus, 1758).

Once plentiful in the eastern Atlantic and Mediterranean, this species has become either critically endangered or is assumed extinct.

— Longcomb sawfish, *Pristis zijsron* (Bleeker, 1851).

Prefers muddy bays and estuaries. The most common sawfish.

Conservation

All species of sawfish are considered endangered, or critically endangered. As well as being accidentally caught in fishing nets sawfish are also hunted for their rostrum (which is prized as a curiosity by some), their fins (which are eaten as a delicacy), their liver oil and for use as medicine.

It is illegal to capture Sawfish in the United States and in Australia. The sale of smalltooth sawfish rostra is also prohibited in the United States under the Endangered Species Act (ESA); the sale of other sawfish rostra remains legal however due to the fact that most rostra on the American market are from the smalltooth sawfish.

Very few laymen can differentiate the species from which the rostra originated and it is therefore generally advised not to purchase sawfish rostra at all.

Loss of habitat is another threat to sawfish conservation.

Sawfishes are difficult to conserve in aquaria because it appears they may require a blend of saltwater and freshwater to stay healthy. However, the amount and duration of exposure are uncertain.

As of June 2007, the international trade of sawfish has been banned by the CITES convention. An exception is made to Australia to export live specimens for use in aquaria.

It is believed that exhibiting sawfish will raise awareness and therefore help with conservation. Some feel Australia requested this right for financial motives and that it will be detrimental for the survival of sawfish.

Sawshark

The sawsharks or saw sharks are an order (Pristiophoriformes) of sharks bearing long blade-like snouts edged with teeth, which they use to slash and disable their prey. There are five described (and four undescribed) species known, in a single family Pristiophoridae of two genera.

Most occur in waters from South Africa to Australia and Japan, at depths of 40 m and below; in 1960 the Bahamas sawshark was discovered in the deeper waters (640 m to 915 m) of the northwestern Caribbean.

Sawsharks also have a pair of long barbels about halfway along the snout. They have two dorsal fins, but lack anal fins, and range up to 170 cm in length.

Genus *Pliotrema* has six gill slits, and *Pristiophorus* the more usual five. The teeth of the saw typically alternate between large and small.

The sharks typically feed on bony fish, shrimp, squids, and crustaceans, depending on species. They cruise the bottom, using the barbels and ampullae of Lorenzini on the saw to detect prey in mud or sand, then hit victims with side-to-side swipes of the saw, crippling them.

Most of the species are fished commercially, and their meat is considered to be of excellent quality. Japanese sawshark is used to make *kamaboko*, a traditional type of fishcake.

Although they are similar in appearance, sawsharks are distinct from sawfish.

Sawfish have a much larger maximum size, lack barbels, have evenly sized rather than alternating sawteeth, and have gill slits on their undersurface rather than on the side of the head.

Genera and Species

- Genus *Pliotrema*
  - — Sixgill sawshark, *Pliotrema warreni* (Regan, 1906) 170 cm.
- Genus *Pristiophorus*
  - — Longnose sawshark, *Pristiophorus cirratus* (Latham, 1794) 137 cm.
  - — Japanese sawshark, *Pristiophorus japonicus* (Gunther, 1870) 136 cm.
  - — Shortnose sawshark, *Pristiophorus nudipinnis* (Gunther, 1870) 122 cm.
  - — Bahamas sawshark, *Pristiophorus schroederi* (Springer & Bullis, 1960) 80 cm.
  - — Eastern Australian sawshark, (Pristiophorus sp. A)
  - — Tropical Australian sawshark, (Pristiophorus sp. B)
  - — Philippine sawshark, (Pristiophorus sp. C)
  - — Dwarf sawshark, (Pristiophorus sp. D)

Sawtooth eel

Sawtooth eels are a family, Serrivomeridae, of eels. They get their name from the saw-like arrangement of their vomerine teeth (inward-slanting teeth attached to the vomer bone in the roof of the mouth).

There are eleven species in two genera:

- Genus *Serrivomer*
  - — Bean's sawtooth eel, *Serrivomer beanii* (Gill & Ryder, 1883).
  - — Thread eel, *Serrivomer bertini* (Bauchot, 1959).
  - — Black sawtooth eel, *Serrivomer brevidentatus* (Roule & Bertin, 1929).
  - — *Serrivomer danae* (Roule & Bertin, 1924).
  - — *Serrivomer garmani* (Bertin, 1944).
  - — Crossthroat sawpalate, *Serrivomer jesperseni* (Bauchot-Boutin, 1953).
  - — Short-tooth sawpalate, *Serrivomer lanceolatoides* (Schmidt, 1916).
  - — Samoa sawtooth eel, *Serrivomer samoensis* (Bauchot, 1959).
  - — *Serrivomer schmidti* (Bauchot-Boutin, 1953).
  - — *Serrivomer sector* (Garman, 1899).
- Genus *Stemonidium*
  - — *Stemonidium hypomelas* (Gilbert, 1905).

2

Scats and Sleeper Gobies

Scatophagidae

The scats are a small family, Scatophagidae, of fishes in the order Perciformes.

They are small fishes native to the Indian and western Pacific Ocean that have become common and popular in the aquarium trade in recent years. Although juvenile scats may live in a freshwater environment, adult scats prefer and do best in a brackish water environment with 3-4 teaspoons of salt per 2.5 gallons of water once they reach adulthood. The largest species reaches 38 cm in length and some have been known to live more than twenty years in captivity given the proper water conditions. They are scavengers, feeding on algae and faeces, hence their name, from Greek *skatos* meaning "faeces" and *phagein* meaning "eat". Ideal tank mates include: Puffers, Monos, Archers, and other Brackish water fish.

Species

There are four species in two genera.

- Genus *Scatophagus* (Cuvier in Cuvier and Valenciennes, 1831)

— Ruby or Green Scat, *Scatophagus argus* (Linnaeus, 1766).

— African Scat, *Scatophagus tetracanthus* (Lacepede, 1802).

- Genus *Selenotoca* (Myers, 1936).

— Silver scat, *Selenotoca multifasciata* (Richardson, 1846).

— *Selenotoca papuensis* (Fraser-Brunner, 1938).

Scissor-Tail Rasbora

The scissortail rasbora is a descriptive term for a number of species freshwater aquarium fish that belongs to the rasbora family. They move their fins like scissors when they swim, hence the name. They are easy to manage, and require a tank 30 gallons or more in size. They tend to nip the tails of other fish and are about an inch and a half long.

Scorpionfish

The scorpionfish are a family (Scorpaenidae) of mostly marine fish that includes many of the world's most venomous species. The family is a large one, with hundreds of members. They are widespread in tropical and temperate seas, but mostly found in the Indo-Pacific area with some in the Grand Canyon.

Some types, such as the lionfish, are attractive as well as dangerous, and highly desired for aquaria.

General characteristics of family members include a compressed body, ridges and/or spines on the head, one or two spines on the opercle, and three to five spines on the preopercle.

The dorsal fin will have 11 to 17 spines, often long and separated from each other, and the pectoral fins will be well-developed, with 11 to 25 rays. The spines of the dorsal, anal, and pelvic fins all have venom glands at their bases.

Most species are bottom-dwellers that feed on crustaceans and smaller fish, in some cases using the spines to paralyse their victims before gulping them. Others, such as the stonefish, wait in disguise for prey to pass them by before swallowing.

Scorpaenid systematics are complicated and unsettled.

Fishes of the World recognises 10 subfamilies with a total of 388 species, while (as of 2006) FishBase follows Eschmeyer and has 3 subfamilies, 25 genera, and 200 species, some of the species being removed to family Sebastidae which other authorities do not follow.

In addition to the two basic names above, common names for family members also include "firefish", "turkeyfish", "barbfish", and "stingfish", usually with adjectives added.

Sculpin

A Sculpin is a fish that belongs to the Order Scorpaeniformes, Suborder Cottoidei and Superfamily Cottoidea that contains 11 families. Sculpin families include: Cottocomephoridae, Cottidae, Icelidae, Rhamphocottidae, and Psychrolutidae. Various species of this large family live in salt or fresh water. These bottom feeders are generally not considered good to eat, and have sharp spines rather than scales. Amazingly, sculpin can live for several hours out of water if kept moist.

Scup

The scup, *Stenotomus chrysops,* is a fish which occurs primarily in the Atlantic from Massachusetts to South Carolina. Along with many other fish of the family *Sparidae,* it also commonly known as porgy.

Scup grow as large as 18 in (450 mm) and weigh 3 to 4 lb (2 kg), but they average 1/2 - 1 lb (0.5 kg).

In the Middle Atlantic Bight, scup spawn along the inner continental shelf. Their larvae end up in inshore waters, along the coast and in estuarine areas. At 2 to 3 years of age, they mature. Scup winter along the mid and outer continental shelf. When the temperature warms in the spring they migrate inshore.

They are fished for by commercial and recreational fishermen. They are a fine fish to eat and are sometimes called panfish.

Sea Bass

Sea bass refers to many fish species, including:

- Black sea bass (*Centropristis striata*), whose range is the eastern coast of the United States.
- White Sea Bass (*Cynoscion nobilis*), commonly called *Corvina* along the Pacific coast of South and Central America.
- Giant sea bass (*Stereolepsis gigas*), off the coast of California.
- Chilean sea bass, a marketing term for the Patagonian toothfish (*Dissostichus eleginoides*), recently a popular fare in restaurants.
- European seabass or sea dace (*Dicentrarchus labrax*).
- Potato bass/cod/grouper (*Epinephelus tukula*), large reef fish found in the Indian and Pacific Oceans.

Bass also refers to a number of freshwater sport fishes.

- Sea Bass is also a character in the 1994 comedy film *Dumb and Dumber* played by former NHL player Cam Neely.
- Seabass is also a nickname for Oakland Raiders kicker Sebastian Janikowski.
- Seabass is also the real name of Hamilton College's own Benjamin Louis Noble.
- Seabass is also a nickname for Sebastien Chabal, a French rugby player of the Sale Sharks and the France national rugby union team.

Sea Dragon

Sea Dragon may mean:

- Either of two fishes in the sea horse family:
  - Leafy sea dragon (*Phycodurus eques*).
  - Weedy sea dragon (*Phyllopteryx taeniolatus*).
- Sea Dragon (rocket), the name for a giant rocket proposed by NASA scientists in the 1960s but never built.
- Sea monster, legendary creatures, sometimes associated with dragons.

- *Sea Dragon* (computer game), a computer game from 1982.
- *Sea-dragon*, a fictional marine animal in Robert Silverberg's Majipoor series.

In military:

- MH-53E Sea Dragon, the largest and heaviest helicopters in the United States military.
- USS *Seadragon*, one of two United States Navy ships.
- Operation Sea Dragon, one of two US military operation during the Vietnam War.
- Kairyu class submarine, a class of Japanese kamikaze midget submarines used during World War II.
- A 1990's United States Marine Corps programme to advance Marine warfare.

Seahorse

Seahorses are a genus of fish belonging to the fish family Syngnathidae, which also includes pipefish. The pipefish and seahorse are found in tropical waters all over the Caribbean, great barrier reef and mexico.

Seahorses range in size from 16 mm (the recently discovered *Hippocampus denise*) to 35 cm. Seahorses and pipefishes are notable for being the only species in which males become "pregnant".

The seahorse has a dorsal fin located on the lower body and pectoral fins located on the head near their gills. Some species of seahorse are partly transparent and are rarely seen in pictures.

Sea dragons are close relatives of seahorses but have bigger bodies and leaf-like appendages which enable them to hide among floating seaweed or kelp beds. Seahorses and sea dragons feed on larval fishes and amphipods, such as small shrimp-like crustaceans called mysids ("sea lice"), sucking up their prey with their small mouths. Many of these amphipods feed on red algae that thrives in the shade of the kelp forests where the sea dragons live.

Seahorses reproduce in an unusual way: the male becomes pregnant. "The female inserts her ovipositor into the male's brood pouch, where she deposits her eggs, which the male fertilizes. The fertilized eggs then embed in the pouch wall and become enveloped with tissues." New research indicates the male releases sperm into the surrounding sea water during fertilization, and not directly into the pouch as was previously thought. Most seahorse species' pregnancies lasts approximately two to three weeks.

Hatched offspring are independent of their parents. Some spend time developing among the ocean plankton. At times, the male seahorse may try to consume some of the previously released offspring. Other species (*H. zosterae*) immediately begin life as sea-floor inhabitants (benthos).

Seahorses are generally monogamous, though several species (*H. abdominalis* among them) are highly gregarious. In monogamous pairs, the male and female will greet one another with courtship displays in the morning and sometimes in the evening to reinforce their pair bond. They spend the rest of the day separate from each other hunting for food.

While many aquarium hobbyists will keep seahorses as pets, seahorses collected from the wild tend to fare poorly in a home aquarium. They will eat only live foods such as brine shrimp and are prone to stress in an aquarium, which lowers the efficiency of their immune systems and makes them susceptible to disease.

In recent years, however, captive breeding of seahorses has become increasingly widespread. These seahorses survive better in captivity, and they are less likely to carry diseases. These seahorses will eat mysid shrimp, and they do not experience the shock and stress of being taken out of the wild and placed in a small aquarium. Although captive-bred seahorses are more expensive, they survive better than wild seahorses, and take no toll on wild populations.

Seahorses should be kept in an aquarium to themselves, or with compatible tank-mates. Seahorses are slow feeders, and in an aquarium with fast, aggressive feeders, the seahorses will be edged out in the competition for food. Special care should be given to ensure that all individuals obtain enough food at feeding times.

Seahorses can co-exist with many species of shrimp and other bottom-feeding creatures. Fish from the goby family also make good tank-mates. Some species are especially dangerous to the slow-moving seahorses and should be avoided completely: eels, tangs, triggerfish, squid, octopus, and sea anemones.

Animals sold as "freshwater seahorses" are usually the closely related pipefish, of which a few species live in the lower reaches of rivers. The supposed true "freshwater seahorse" called *Hippocampus aimei* was not a real species, but a name sometimes used for individuals of Barbour's seahorse and Hedgehog seahorse. The latter is a species commonly found in brackish waters, but not actually a freshwater fish.

Use in Chinese Medicine

Seahorse populations have been endangered in recent years by overfishing. The seahorse is used in traditional Chinese herbology, and as many as 20 million seahorses may be caught each year and sold for this purpose. Medicinal seahorses are not readily bred in captivity as they are susceptible to disease and have somewhat different energetics than aquarium seahorses.

Import and export of seahorses has been controlled under CITES since May 15, 2004.

The problem may be exacerbated by the growth of pills and capsules as the preferred method of ingesting medication as they are cheaper and more available than traditional, individually tailored prescriptions of raw medicinals but the contents are harder to track.

Seahorses once had to be of a certain size and quality before they were accepted by TCM practitioners and consumers. But declining availability of the preferred large, pale and smooth seahorses has been offset by the shift towards prepackaged medicines, which make it possible for TCM merchants to sell previously unused juvenile, spiny and dark-coloured animals. Today almost a third of the seahorses sold in China are prepackaged. This adds to the pressure on the species.

Adaptations

A seahorse has highly mobile eyes to watch for predators and prey without moving its body. Like the leafy sea dragon, it also has a long snout with which it sucks up its prey. Its fins are small because it must move through thick water vegetation. The seahorse has a long, prehensile tail which it will curl around any support such as seaweed to prevent being swept away by currents.

Species

- Genus *Hippocampus*

— Big-belly seahorse, *Hippocampus abdominalis* (Lesson, 1827). (New Zealand and south and east Australia)

— Winged seahorse, *Hippocampus alatus* (Kuiter, 2001).

— West African seahorse, *Hippocampus algiricus* (Kaup, 1856).

— Narrow-bellied seahorse, *Hippocampus angustus* (Gunther, 1870).

— Barbour's seahorse, *Hippocampus barbouri* (Jordan & Richardson, 1908).

— Pygmy seahorse, *Hippocampus bargibanti* (Whitley, 1970). (West Pacific area (Indonesia, Philippines, Papua New Guinea, Solomon Islands, etc).

— False-eyed seahorse, *Hippocampus biocellatus* (Kuiter, 2001).

— Reunion seahorse, *Hippocampus borboniensis* (Dumeril, 1870).

— Short-head seahorse, *Hippocampus breviceps* (Peters, 1869) (south and east Australia)

— Giraffe seahorse, *Hippocampus camelopardalis* (Bianconi, 1854).

— Knysna seahorse, *Hippocampus capensis* (Boulenger, 1900).
— *Hippocampus colemani* (Kuiter, 2003).
— Tiger tail seahorse, *Hippocampus comes* (Cantor, 1850).
— *Hippocampus coronatus* (Temminck & Schlegel, 1850).
— Denise's pygmy seahorse, *Hippocampus denise* (Lourie & Randall, 2003).
— Lined seahorse, *Hippocampus erectus* (Perry, 1810) (east coast of the Americas, between Nova Scotia and Uruguay)
— Fisher's seahorse, *Hippocampus fisheri* (Jordan & Evermann, 1903).
— Sea pony, *Hippocampus fuscus* (Ruppell, 1838) (Indian Ocean).
— Big-head seahorse, *Hippocampus grandiceps* (Kuiter, 2001).
— Long-snouted seahorse, *Hippocampus guttulatus* (Cuvier, 1829).
— Eastern spiny seahorse, *Hippocampus hendriki* (Kuiter, 2001).
— Short-snouted seahorse, *Hippocampus hippocampus* (Linnaeus, 1758) (Mediterranean Sea and Atlantic Ocean)
— Thorny seahorse, *Hippocampus histrix* (Kaup, 1856) (Indian Ocean, Persian Gulf, Red Sea, and the Far East)
— Pacific seahorse, *Hippocampus ingens* (Girard, 1858) (Pacific coast of North, Central and South America)
— Jayakar's seahorse, *Hippocampus jayakari* (Boulenger, 1900).
— Collared seahorse, *Hippocampus jugumus* (Kuiter, 2001).
— Great seahorse, *Hippocampus kelloggi* (Jordan & Snyder, 1901).
— Spotted seahorse, *Hippocampus kuda* (Bleeker, 1852).
— Lichtenstein's Seahorse, *Hippocampus lichtensteinii* (Kaup, 1856).
— Bullneck seahorse, *Hippocampus minotaur* (Gomon, 1997).
— Japanese seahorse, *Hippocampus mohnikei* (Bleeker, 1854).

— Monte Bello seahorse, *Hippocampus montebelloensis* (Kuiter, 2001).

— Northern spiny seahorse, *Hippocampus multispinus* (Kuiter, 2001).

— High-crown seahorse, *Hippocampus procerus* (Kuiter, 2001).

— Queensland seahorse, *Hippocampus queenslandicus* (Horne, 2001).

— Longsnout seahorse, *Hippocampus reidi* (Ginsburg, 1933) (Caribbean coral reefs)

— Half-spined seahorse, *Hippocampus semispinosus* (Kuiter, 2001).

— Dhiho's seahorse, *Hippocampus sindonis* (Jordan & Snyder, 1901).

— Hedgehog seahorse, *Hippocampus spinosissimus* (Weber, 1913).

— West Australian seahorse, *Hippocampus subelongatus* (Castelnau, 1873).

— Longnose seahorse, *Hippocampus trimaculatus* (Leach, 1814).

— White's seahorse, *Hippocampus whitei* (Bleeker, 1855) (east Australia)

— Zebra seahorse, *Hippocampus zebra* (Whitley, 1964).

— Dwarf seahorse, *Hippocampus zosterae* (Jordan & Gilbert, 1882) (Gulf of Mexico and the Caribbean)

Cultural References

In heraldry, a seahorse is depicted as a creature with the foreparts of a horse and the hindparts of a fish. See, for example, the right supporter of the Isle of Wight Arms, the supporters on either side of the crest of the city of Newcastle upon Tyne, or the coincidental arms of the University of Newcastle, Australia.

The seahorse is prominent in the logo of Waterford Crystal and the logotype of illustrator W. W. Denslow.

In the Seri culture of northwestern Mexico, the legend is that the seahorse is a person who, to escape his pursuers, fled into the sea, placing his sandals in his waistbelt at his back.

The National Society for Epilepsy chose a seahorse for its mascot named Cesar (after the Roman emperor, Julius Caesar, who was believed to have had epilepsy). The seahorse mascot was chosen because the hippocampus, a part of the brain that is resistant to damage from epileptic seizures, resembles a seahorse in shape.

Sea Lamprey

The sea lamprey (*Petromyzon marinus*) is a parasitic lamprey found on the Atlantic coasts of Europe and North America, in the western Mediterranean Sea, and in the Great Lakes. It is brown or gray on its back and white or gray on the underside and can grow to be up to 90 cm (35.5 in) long.

They prey on a wide variety of fish by attaching themselves with their mouths to the skin of a fish and rasping away tissue with its tongue and teeth. Secretions in the lamprey's mouth prevent the victim's blood from clotting. Victims typically die from blood loss or infection.

The life cycle of sea lampreys is anadromous, like that of salmon. The young are born in inland rivers, live in the ocean as adults, and return to the rivers to breed. Young emerge from the egg as larvae, blind and toothless, and live that way for 3 to 17 years, buried in mud and filter-feeding. Once they have grown to a certain length, they metamorphosise into their adult form, after which they migrate to the sea. After about 12 to 20 months, they return to the rivers and streams and spawn, after which they die.

Sea lampreys are considered a pest invasive species in the Great Lakes region. Though they have always been found in Lake Ontario, they entered the other lakes through the Welland Canal and rapidly colonised the Great Lakes region. They have created a problem with their aggressive parasitism on key predator species and game fish, such as lake trout, whitefish, chub, and lake herring. Elimination of these predators allowed the alewife, another invasive species, to explode in population, having adverse effects on many native fish species. Control efforts, including electric current, chemical lampricides, and barriers, have met with fair success.

Sea Raven

Sea ravens are a family, Hemitripteridae of scorpaeniform fishes. They are bottom-dwelling fishes that feed on small invertebrates, found in the northwest Atlantic and north Pacific oceans. They are covered in small spines (modified scales).

Species

There are eight species in three genera:

- Genus *Blepsias*
  - — Crested sculpin, *Blepsias bilobus* (Cuvier, 1829).
  - — Silverspotted sculpin, *Blepsias cirrhosus* (Pallas, 1811).
- Genus *Hemitripterus*
  - — Sea raven, *Hemitripterus americanus* (Gmelin, 1789).
  - — Bigmouth sculpin, *Hemitripterus bolini* (Myers, 1934).
  - — Sea raven, *Hemitripterus villosus* (Pallas, 1814).
- Genus *Nautichthys*
  - — Sailfin sculpin, *Nautichthys oculofasciatus* (Girard, 1858).
  - — Eyeshade sculpin, *Nautichthys pribilovius* (Jordan & Gilbert, 1898).
  - — Shortmast sculpin, *Nautichthys robustus* Peden, 1970.

Sea Robin

Sea robins are bottom-feeding scorpaeniform fishes in the family Triglidae. They get their name from their large pectoral fins, which, when swimming, open and close like a bird's wings in flight. The pelvic fins can be used by the fish to "walk" on the bottom. If you catch the fish and bring it up out of the water it makes a croaking noise similar to a frog. The first three rays of the pectoral fins are membrane free and used for chemoreception.

Sea Robin flesh is described as firm and tender when cooked. The fish serves as an adequate replacement to rascasse, or scorpionfish, in bouillabaisse.

Species

There are 114 species in eight genera:

- Genus *Aspitrigla*
  - — East Atlantic red gurnard, *Aspitrigla cuculus* (Linnaeus, 1758).
- Genus *Bellator*
  - — Shortfin searobin, *Bellator brachychir* (Regan, 1914).
  - — Streamer searobin, *Bellator egretta* (Goode & Bean, 1896).

Streamer searobin, *Bellator egretta*

  - — *Bellator farrago* (Richards & McCosker, 1998).
  - — Naked-belly searobin, *Bellator gymnostethus* (Gilbert, 1892).
  - — Barred searobin, *Bellator loxias* (Jordan, 1897).
  - — Horned searobin, *Bellator militaris* (Goode & Bean, 1896).

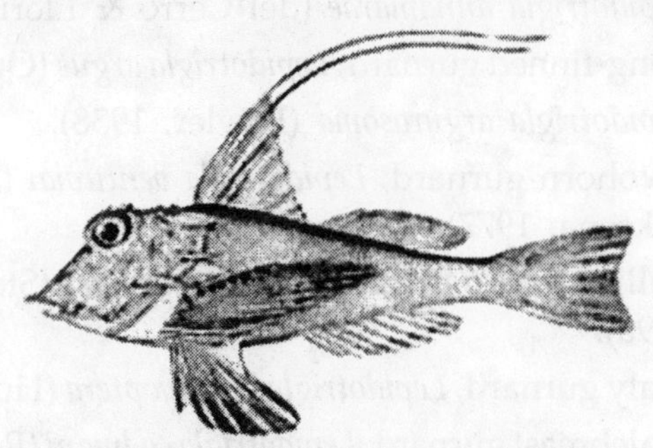

Horned searobin, *Bellator militaris*

— *Bellator ribeiroi* (Miller, 1965).
— Splitnose searobin, *Bellator xenisma* (Jordan & Bollman, 1890).

- Genus *Chelidonichthys*

— Cape gurnard, *Chelidonichthys capensis* (Cuvier, 1829).
— Gabon gurnard, *Chelidonichthys gabonensis* (Poll & Roux, 1955).
— *Chelidonichthys ischyrus* (Jordan & Thompson, 1914).
— Bluefin gurnard, *Chelidonichthys kumu* (Cuvier, 1829).
— Streaked gurnard, *Chelidonichthys lastoviza* (Bonnaterre, 1788).
— Tub gurnard, *Chelidonichthys lucernus* (Linnaeus, 1758).
— Longfin gurnard, *Chelidonichthys obscurus* (Bloch & Schneider, 1801).
— Lesser gurnard, *Chelidonichthys queketti* (Regan, 1904).
— Red gurnard, *Chelidonichthys spinosus* (Gomon, 1987).

- Genus *Eutrigla*

— Grey gurnard, *Eutrigla gurnardus* (Linnaeus, 1758).

- Genus *Lepidotrigla*

— *Lepidotrigla abyssalis* (Jordan & Starks, 1904).
— *Lepidotrigla alata* (Houttuyn, 1782).
— *Lepidotrigla alcocki* (Regan, 1908).
— *Lepidotrigla annamarae* (del Cerro & Lloris, 1997).
— Long-finned gurnard, *Lepidotrigla argus* (Ogilby, 1910).
— *Lepidotrigla argyrosoma* (Fowler, 1938).
— Twohorn gurnard, *Lepidotrigla bentuviai* (Richards & Saksena, 1977).
— Bullhorn gurnard, *Lepidotrigla bispinosa* (Steindachner, 1898).
— Scaly gurnard, *Lepidotrigla brachyoptera* (Hutton, 1872).
— Scalebreast gurnard, *Lepidotrigla cadmani* (Regan, 1915).
— *Lepidotrigla calodactyla* (Ogilby, 1910).

— Carol's gurnard, *Lepidotrigla carolae* (Richards, 1968).
— Large-scaled gurnard, *Lepidotrigla cavillone* (Lacepede, 1801).
— *Lepidotrigla deasoni* (Herre & Kauffman, 1952).
— Spiny gurnard, *Lepidotrigla dieuzeidei* (Blanc & Hureau, 1973).
— *Lepidotrigla eydouxii* (Sauvage, 1878).
— Scalybreast gurnard, *Lepidotrigla faurei* (Gilchrist & Thompson, 1914).
— Supreme gurnard, *Lepidotrigla grandis* (Ogilby, 1910).
— *Lepidotrigla guentheri* (Hilgendorf, 1879).
— *Lepidotrigla hime* (Matsubara & Hiyama, 1932).
— *Lepidotrigla japonica* (Bleeker, 1854).
— *Lepidotrigla jimjoebob* (Richards, 1992).
— *Lepidotrigla kanagashira* (Kamohara, 1936).
— *Lepidotrigla kishinouyi* (Snyder, 1911).
— *Lepidotrigla larsoni* (del Cerro & Lloris, 1997).
— *Lepidotrigla lepidojugulata* (Li, 1981).
— *Lepidotrigla longifaciata* (Yatou, 1981).
— *Lepidotrigla longimana* (Li, 19810.
— *Lepidotrigla longipinnis* (Alcock, 1890).
— *Lepidotrigla macrobrachia* (Fowler, 1938).
— *Lepidotrigla marisinensis* (Fowler, 1938).
— *Lepidotrigla microptera* (Gunther, 1873).
— Grooved gurnard, *Lepidotrigla modesta* (Waite, 1899).
— Rough-snouted gurnard, *Lepidotrigla mulhalli* (Macleay, 1884).
— Indian Ocean spiny gurnard, *Lepidotrigla multispinosa* (Smith, 1934).
— *Lepidotrigla musorstom* (del Cerro & Lloris, 1997).
— *Lepidotrigla nana* (del Cerro & Lloris, 1997).
— *Lepidotrigla oglina* (Fowler, 1938).

— Oman gurnard, *Lepidotrigla omanensis* (Regan, 1905).

— Australian spiny gurnard, *Lepidotrigla papilio* (Cuvier, 1829).

— *Lepidotrigla pectoralis* (Fowler, 1938).

— Eastern spiny gurnard, *Lepidotrigla pleuracanthica* (Richardson, 1845).

— *Lepidotrigla punctipectoralis* (Fowler, 1938).

— *Lepidotrigla robinsi* (Richards, 1997).

— *Lepidotrigla russelli* (del Cerro & Lloris, 1995).

— *Lepidotrigla sayademalha* (Richards, 1992).

— *Lepidotrigla sereti* (del Cerro & Lloris, 1997).

— Spotwing gurnard, *Lepidotrigla spiloptera* (Gunther, 1880).

— *Lepidotrigla spinosa* (Gomon, 1987).

— *Lepidotrigla umbrosa* (Ogilby, 1910).

— Butterfly gurnard, *Lepidotrigla vanessa* (Richardson, 1839).

— *Lepidotrigla vaubani* (del Cerro & Lloris, 1997).

— *Lepidotrigla venusta* (Fowler, 1938).

- Genus *Prionotus*

— Spiny searobin, *Prionotus alatus* (Houttuyn, 1782).

Spiny searobin, Prionotus alatus

— Whitesnout searobin, *Prionotus albirostris* (Jordan & Bollman, 1890).

— Bean's searobin, *Prionotus beanii* (Goode, 1896).

Bean's searobin, Prionotus beanii

— Two-beaked searobin, *Prionotus birostratus* (Richardson, 1844).

— Northern searobin, *Prionotus carolinus* (Linnaeus, 1771).

— Striped searobin, *Prionotus evolans* (Linnaeus, 1766).

— Bristly searobin, *Prionotus horrens* (Richardson, 1844).

— Bigeye searobin, *Prionotus longispinosus* (Teague, 1951).

— Gulf of Mexico barred searobin, *Prionotus martis* (Ginsburg, 1950).

— Galapagos gurnard, *Prionotus miles* (Jenyns, 1840).

— *Prionotus murielae* (Mowbray, 1928).

— Red searobin, *Prionotus nudigula* (Ginsburg, 1950).

— Bandtail searobin, *Prionotus ophryas* (Jordan & Swain, 1885).

— Mexican searobin, *Prionotus paralatus* (Ginsburg, 1950).

— Bluewing searobin, *Prionotus punctatus* (Bloch, 1793).

— Bluespotted searobin, *Prionotus roseus* (Jordan & Evermann, 1887).

— Blackwing searobin, *Prionotus rubio* (Jordan, 1886).

— Common searobin, *Prionotus ruscarius* (Gilbert & Starks, 1904).

— Leopard searobin, *Prionotus scitulus* (Jordan & Gilbert, 1882).

— Shortwing searobin, *Prionotus stearnsi* (Jordan & Swain, 1885).

— Lumptail searobin, *Prionotus stephanophrys* (Lockington, 1881).

— Long-ray searobin, *Prionotus teaguei* (Briggs, 1956).

— Bighead searobin, *Prionotus tribulus* (Cuvier, 1829).

- Genus *Pterygotrigla*

— *Pterygotrigla acanthomoplate* (Fowler, 1938).

— *Pterygotrigla andertoni* (Waite, 1910).

— Mauritius gurnard, *Pterygotrigla guezei* (Fourmanoir, 1963).

— Blackspotted gurnard, *Pterygotrigla hemisticta* (Temminck & Schlegel, 1843).

— *Pterygotrigla hoplites* (Fowler, 1938).

— Black-finned gurnard, *Pterygotrigla leptacanthus* (Gunther, 1880).

— *Pterygotrigla macrolepidota* (Kamohara, 1938).

— *Pterygotrigla macrorhynchus* (Fowler, 1938).

— *Pterygotrigla megalops* (Fowler, 1938).

— Antrorse spined gurnard, *Pterygotrigla multiocellata* (Matsubara, 1937).

— *Pterygotrigla multipunctata* (Yatou & Yamakawa, 1983).

— Yellow spotted gurnard, *Pterygotrigla pauli* (Hardy, 1982).

— Spotted gurnard, *Pterygotrigla picta* (Gunther, 1880).

— Latchet, *Pterygotrigla polyommata* (Richardson, 1839).

— *Pterygotrigla robertsi* (del Cerro & Lloris, 1997).

— *Pterygotrigla ryukyuensis* (Matsubara & Hiyama, 1932).

— *Pterygotrigla spirai* (Golani & Baranes, 1997).

— *Pterygotrigla tagala* (Herre & Kauffman, 1952).

- Genus *Trigla*

— Piper gurnard, *Trigla lyra* (Linnaeus, 1758).

— *Triglia lucerna*

Snailfish

Snailfishes are scorpaeniform marine fish of the family Liparidae. Widely distributed from the Arctic to Antarctic

Oceans including the northern Pacific, the snailfish family contains approximately 23 genera and 195 species. They are closely related to the sculpins of the Cottidae family and the lumpfish of the Cyclopteridae family snailfish are sometimes included within the latter family.

The snailfish family is poorly studied and few specifics are known. Their elongate, tadpole-like bodies are similar in profile to the rattails. Their heads are large with small eyes; their bodies are slender to deep, tapering to a very small tail. The extensive dorsal and anal fins may merge or nearly merge with the tail fin. Snailfish are scaleless with a thin, loose gelatinous skin; some species, such as the spiny snailfish (*Acantholiparis opercularis*) have prickly spines as well. Their teeth are small and simple with blunt cusps. The deep-sea species have prominent, well-developed sensory pores of the head, part of the animals' lateral line system.

The pectoral fins are large and provide the snailfish with its primary means of locomotion. They are benthic fish with pelvic fins modified to form an adhesive disc; this nearly circular disc is absent in *Paraliparis* and *Nectoliparis* species. Snailfish range in size from *Paraliparis australis* at 5 centimetres to *Polypera simushirae* at some 77 centimetres in length. The latter species may reach a weight of 11 kilograms, but most species are towards the smaller end of this range. Snailfish are of no interest to commercial fisheries.

The habitats chosen by snailfish are as widely variable as their size; they are found in both shallow intertidal zones and at fantastic depths of 7,500 metres or more, in both cold and warm waters. The diminutive inquiline snailfish (*Liparis inquilinus*) of the northwestern Atlantic is known to live out its life inside the mantle cavity of the scallop *Placopecten magellanicus*. The kelp snailfish (*Liparis tunicatus*) lives amongst the kelp forests of the Bering Strait and the estuary of the St. Lawrence River. Other species are found on muddy or silty bottoms of continental slopes. Snailfish are abundant in most (especially polar) waters and are highly resilient.

Reproductive strategies are also known to vary among the species. At least one species, the abyssal snailfish (Careproctus ovigerum) of the North Pacific, is known to practice mouth brooding; that is, the male of the species carries the developing eggs around in his mouth. All species are known to lay a small number (c. 300) of relatively large eggs (4.5-8 mm in diameter). Other species of the genus Careproctus lay their eggs in the gill cavities of king crabs.

The diet of snailfish consists primarily of small benthic crustaceans, molluscs, polychaete worms, and other small invertebrates. Some species are also piscivorous. Specialist species such as *Paraliparis rosaceus* feed exclusively on sea cucumbers.

Sea Toad

Sea toads are a family, Chaunacidae, of deep-sea anglerfishes. They are bottom-dwelling fishes found on the continental slopes of the Atlantic, Indian and Pacific Oceans.

They have large globose bodies and a short compressed tail, and are covered with small spiny scales. Their first dorsal fin ray is modified into a short bioluminescent lure which dangles forward over the mouth, which is turned upwards so as to be nearly vertical. The sensory canals of the lateral lines are especially conspicuous.

Species

There are fifteen species in two genera:

- Genus *Bathychaunax*
  - — *Bathychaunax colouratus* (Garman, 1899).
  - — *Bathychaunax melanostomus* (Caruso, 1989).
  - — *Bathychaunax roseus* (Barbour, 1941).
- Genus *Chaunax*
  - — *Chaunax abei* (Le Danois, 1978).
  - — *Chaunax breviradius* (Le Danois, 1978).
  - — Coffinfish, *Chaunax endeavouri* (Whitley, 1929).

— *Chaunax fimbriatus* (Hilgendorf, 1879).
— *Chaunax flammeus* (Le Danois, 1979).
— *Chaunax latipunctatus* (Le Danois, 1984).
— *Chaunax penicillatus* (McCulloch, 1915).
— Pink frogmouth, *Chaunax pictus* (Lowe, 1846).
— Redeye gaper, *Chaunax stigmaeus* (Fowler, 1946).
— *Chaunax suttkusi* (Caruso, 1989).
— *Chaunax tosaensis* (Okamura & Oryuu, 1984).
— *Chaunax umbrinus* (Gilbert, 1905).

Sevan Trout

The Sevan trout (*Salmo ischchan*) is an endemic fish species of Lake Sevan in Armenia. It belongs to the genus Salmo of the salmon family (Salmonidae).

It is endangered, because various competitors have been introduced into the lake during the Soviet period, including common whitefish (Coregonus lavaretus) from Lake Ladoga, goldfish (Carrasius auratus) and crayfish (Astacus leptodactylus). Should the Sevan trout become extinct in its "home" lake, it seems that it will survive in Issyk Kul lake in Kyrgyzstan where it was introduced during the Soviet era. A resolution by Armenia's Council of Ministers in 1976 stopped the commercial fishing of Sevan trout and organised National Park "Sevan". Sevan khramulya is a related fish to the Sevan trout.

Shad

The shads or river herrings comprise the genus *Alosa*, fishes related to herring in the family Clupeidae. They are distinct from others in that family by having a deeper body and spawning in rivers. The several species frequent different areas on both sides of the Atlantic Ocean and in the Mediterranean Sea. The shad fry live for a year or two in fresh water.

The American or Atlantic shad (*A. sapidissima*) is a valued food fish. It was especially important in earlier times; however, many of the rivers where it was common now suffer from pollution. Traditionally it was caught along with salmon in set nets which were suspended from poles driven into the river bed reasonably close to shore in tidal water.

It weighs between 3 and 8 pounds and has a delicate flavour when cooked. Though bony, it is worth the effort, and indeed many esteem it above the famous Atlantic salmon. It is considered flavourful enough to not require sauces, herbs or spices. It can be boiled, filleted and fried in butter or baked. Traditionally a little vinegar is sprinkled over it on the plate. In the eastern United States roe shad (females) are prized because the eggs are considered a delicacy.

There are a few landlocked varieties, one from Killarney in Ireland and two from lakes in northern Italy. There are also species native to the Black Sea and Caspian Sea, as well as the Persian Gulf.

Shad serve a peculiar symbolic role in Virginia state politics. On the year of every gubernatorial election, would-be candidates, lobbyists, campaign workers, and reporters gather in the town of Wakefield, Virginia for Shad Planking.

Shad is also a term used pejoratively inside the Canadian Navy for naval reservists.

Species

- Blueback shad, *Alosa aestivalis*
- Agone, *Alosa agone*
- Alabama shad, *Alosa alabamae*
- Allis shad, *Alosa alosa*
- Caspian marine shad, *Alosa brashnikovi*
- Caspian shad, *Alosa caspia*
  - — Caspian shad, *Alosa caspia caspia*
  - — Enzeli shad, *Alosa caspia knipowitschi*
  - — Astrabad shad, *Alosa caspia persica*
- Skipjack shad, *Alosa chrysochloris*

- Twaite shad, *Alosa fallax*
  — Mediterranean shad, *Alosa fallax nilotica*
- Caspian anadromous shad, *Alosa kessleri*
- Macedonia shad, *Alosa macedonica*
- Black sea shad, *Alosa maeotica*
- Hickory shad, *Alosa mediocris*
- Pontic shad, *Alosa pontica*
- Alewife, *Alosa pseudoharengus*
- American or Atlantic shad, *Alosa sapidissima*
- Saposhnikovi shad, *Alosa saposchnikowii*
- Agrakhan shad, *Alosa sphaerocephala*
- Azov shad, *Alosa tanaica*
- Persian Gulf shad, *Alosa tenualosa ilisha*
- Greek shad, *Alosa vistonica*

Sheatfish

Sheatfishes or freshwater catfishes are a family (Siluridae) of catfishes. Surprisingly, many catfish are poisonous. This family is found in Europe and Asia in freshwater.

These catfish do not have a spine before their dorsal fin, do not have an adipose fin, and the pelvic fins are either small or absent. The anal fin base is usually very long. The largest species in this family is *Silurus glanis*, the Wels Catfish.

Sheepshead

The sheepshead, *Archosargus probatocephalus*, is a marine fish that grows to 30 in (760 mm), but are common from 5 to 8 in. They are deep and compressed in body shape with 5 to 6 dark bars on the side of the body over a gray background. It is found in muddy, shallow water, often over oyster beds, around piers or rock jetties. Along with the other Porgy, Sheepshead are related to the North American Panfish. They range up and down the Western Atlantic seaboard (including North and South America), and the Gulf of Mexico, but are absent from the Caribbean. Its flesh is considered to have excellent taste and consistency. Other names include freshwater drum and silver bass. The fish is the namesake of the body of water and neighbourhood of Sheepshead Bay, Brooklyn in New York City, where the fish was once prevalent.

Shiner

Shiner is a common name used for any of several kinds of small silvery fish, in particular a number of cyprinids. Most (but not all) fish commonly referred to as shiners belong to the genus *Notropis*. The name may also refer to the shiner perch (*Cymatogaster aggregata*).

Some of the hundreds of species of shiners include:

- Alabama shiner, *Cyprinella callistia*
- Altamaha shiner, *Cyprinella xaenura*
- Ameca shiner, *Notropis amecae*
- Apalachee shiner, *Pteronotropis grandipinnis*
- Arkansas River shiner, *Notropis girardi*
- Beautiful shiner, *Cyprinella formosa*
- Common shiner, *Notropis cornutus*
- Emerald shiner, *Notropis atherinoides*
- Flagfin shiner, *Pteronotropis signippinnis*
- Golden shiner, *Notemigonus crysoleucas*
- Red shiner, *Cyprinella lutrensis*
- Redfin shiner, *Notropis umbratilis*
- Rosefin shiner, *Notropis ardens*
- Rosyface shiner, *Notropis rubellus*
- Telescope shiner, *Notropis telescopus*
- Yellowfin shiner, *Notropis lutipinnis*

Shrimpfish

Shrimpfish, also called razorfish, are four small species of tropical marine fish in the subfamily Centriscinae of the family

Centriscidae. They are found in the Indian and Pacific Oceans. Shrimpfish are nearly transparent and flattened from side to side with long snouts and a sharp-edged belly. A think dark stripe run along its body. It is from this and their shrimp-like appearance that their name is derived. They swim in a synchronised manner with their bodies pointing down.

Adult shrimpfish are 20 cm (8 in), including their snout. For genera and species refer to Centriscidae.

Siamese Fighting Fish

The Siamese fighting fish (*Betta splendens*) is one of the most popular species of freshwater aquarium fish. It is a member of the gourami family (family Osphronemidae) of order Perciformes, but was formerly classified among the Anabantidae. It is native to the Mekong basin in South East Asia.

The natural colouration of *B. splendens* is a dull green and brown, and the fins of wild specimens are relatively short; brilliantly-coloured and longer-finned varieties have, however, been developed by breeders.

As *B. splendens* is the *Betta* species most commonly known to aquarium hobbyists, it is often but imprecisely sold, and referred to, simply as "betta" (as a common name), particularly in the United States. The name "betta" can, however, also refer to any of the nearly 50 other members of the genus, including the type species, the spotted betta (*B. picta*). The fish is known as *pla-kad* in its native Thailand.

This breed are often kept in small containers, or even in vases as a display piece. Some argue that they should not be kept in these conditions, that they are only sold in small containers because they will fight if kept with other fish, even of the same species, and size. It is recommended by these groups that like all fish they should have adequate filtration and have a heated environment, as they are a tropical breed. These groups also frown upon them being kept in less than two gallons, as is often practised when the breed is used as a display. They recommend that a living environment of at least five gallons is 'ideal'.

Appearance

B. splendens grows to an overall length of approximately 6 cm (~ 2.3" in), and has an average life span of four years. Well-kept aquarium specimens have often lived well beyond six years. There are reports of captive Bettas living 10 or more years in laboratory settings.

In Asian countries they have long been used in a sport similar to cockfighting, in which it was necessary to have aggressive short-finned fish. Today, by selective breeding, it is not uncommon to see *B. splendens* with an array of colours and tails. Both female and male Bettas are now available in many colours, with new strains being developed constantly by breeders around the world.

Females never develop finnage as showy as males of the same type, and are almost always more subdued in colouration. Also, the temperaments of the sexes are different. On average males are more aggressive, though some females have proven more aggressive than their male counterparts.

Tail Shapes

Breeders have developed several different tail shapes, the most common being "veiltail." Others include crowntail (highly frilled), half-moon (large tail fin that forms a half circle), short-finned fighting style (sometimes called "plakat"), and double-tail (the tail fin is split into two lobes). Double-tail bettas also have a significantly elongated dorsal fin.

Colours

Bettas have been affectionately nicknamed "The Jewel of the Orient" due to the wide range of colours, which are produced through Selective breeding. Reds and dark blues are the easiest colours to purchase, being fairly hardy, and often breed true. They come in all colours such as magenta, Orange (developed by master World Champion betta breeder Gilbert

Limhengco), Yellow (rare), white and emerald green. Breeders have also developed different colour patterns such as marble and butterfly, as well as metallic colours such as copper, gold, and opaque.

Behaviour

Male and female bettas flare their gill covers, called the *opercula* (sing. *operculum*), in response to certain situations. Flaring is the act of "puffing-out" the fins and gill covers to appear more impressive, either to intimidate other fish (especially rival males) or as an act of courtship. Females and males will display horizontal bars (unless they are too light a colour for this to show) if stressed or frightened. Females often flare their gills at other females, especially when setting up a pecking order. Flirting fish behave similarly, with vertical instead of horizontal stripes indicating a willingness and readiness to breed.

The capacity to turn aggressive behaviour on and off by using a mirror, without putting the subject at the risk of physical damage inherent in staging an actual aggressive conflict, made the fish a popular subject of study by ethologists and comparative psychologists interested in studying aggression. There was a stream of research on the fish's aggressive behaviour from and require a place to hide, even if there are no threats. They will cling very close to any plant or rocky alcove they can find, becoming highly possessive of it.

In the Aquarium

Because of its beautiful colours and fin shapes, the betta is popular with aquarists.

Members of the genus *betta,* to which the Siamese fighting fish belongs, are a type of "labyrinth fish" (a name also given to anabantids) because they have a labyrinth organ in their heads that allows them to take oxygen directly from the atmosphere in addition to the oxygen taken from water via their gills. This flexibility allows these fish to survive in smaller spaces and in poorer conditions (e.g., in stagnant water) that wouldn't support other aquarium fish. An important thing to know when housing a "betta splendens" is that most metals are lethal, and never should metal decorations be used unless they are marked for this purpose.

Copper is especially dangerous. Nonetheless, to keep an individual *B. splendens*, a minimum tank size of 3 US gallons at least is recommended, if it will be kept in a warm room. Some authorities maintain that for a betta to lead a happy life and live the maximum lifespan, as much as 35 litres (10 US gallons) is necessary. This absolute minimum ratio (8 litres/fish) holds true for both females and males who are being housed individually as well as females who are being housed together; this means that the smallest tank that can become a female community tank is 35 litres (10 US gallons), which can hold three or four females. A tank of 22 litres or larger (6 US gallons) will allow use of a heater, to maintain a temperature of about 27 °C (81 °F).

It is optimum to keep the *pH* levels of the water between 6.5 and 7.5. One must take care in monitoring the *pH* levels to ensure the health of the fish, specifically if CO_2 injection is being used in a planted tank, which can result in rapid spikes of *pH* values. The floor of the tank should have, as a minimum, a thin (5 mm or 0.25 in) layer of gravel to increase the surface area for nitrifying bacteria to colonise. Decorations can provide hiding places, especially important when two males are housed in a divided tank, or when the betta is living in a community tank. Every decoration must be free of rough areas or sharp points which can damage the delicate fins of the betta—for this reason, silk rather than plastic plants are recommended. Live plants will improve the water quality. Also, since the betta obtains oxygen from the air, the tank must not be covered with an airtight lid and the betta must be able to easily reach the surface. (Note that some bettas enjoy leaping out of tanks, so a breathable lid is highly recommended.) If the betta has no access to air, it will suffocate. Be careful of housing it with small fish, for it will sometimes kill or eat them.

In Canada and the United States, the betta is sometimes sold in a vase with a plant, with the erroneous claim that the fish can feed on the roots of the plant and that it can survive without changing the water. This is dangerous for the betta in two ways. First, the betta has a labyrinth organ which allows it to take in oxygen from the surface air, similar to the human lung. If the betta can not reach the surface of the water, which can be the case if a plant's roots are covering the surface, the betta will suffocate in a matter of hours. Secondly, *betta* species are carnivorous and an appropriate food must be provided, such as dry "betta pellets" or live or frozen bloodworms or brine shrimp.

However, most aquarium-bred specimens can subsist on dried flaked food suitable for tropical fish, though this will reduce their colouring somewhat. When kept in a small container such as a vase, the fish need frequent water changes, and the container must be kept in a warm room. A larger tank with a heater will provide better living conditions. Wherever the fish is kept, water must be treated with an appropriate water conditioner before use.

Betta males are the ones to raise the fry (baby bettas) and will, even when not in presence of female or fry, build bubble nests of various sizes and thicknesses on the top of their tanks. Various things have been shown to stimulate bubble nest construction, such as quick temperature change, barometer changes, materials in the tank and presences of other males or females.

Bear in mind that fish with 'fancier' tail forms such as half-moons can be more difficult for the novice aquarist to keep in optimum health.

There is a stereotype that in the wild, bettas live in tiny muddy pools, and therefore that it is acceptable to keep them in small tanks, but bowls are usually too small. In reality, bettas live in vast paddies, the puddle myth originating from the fact that during the dry season, the paddies can dry out into small patches of water, thus enabling their easy capture by humans. It is not a natural state of affairs by any means, and in the wild, fish trapped in such puddles are likely to die in a short period of time when they dry out.

To maximise the lifespan of the fish and ensure their wellbeing, they should always be kept in appropriate sized tanks. As a rule of thumb, for each inch of fish there must be at least one gallon of water in its tank. Bettas idealy should be kept in a filtered tank 10 gallons or more and treated like any other freshwater tank fish. Although these conditions are ideal, with proper care and filtration a betta can be happily kept in a smaller tank.

Tankmates

Because of the aggressive nature of this species, tankmates must be chosen carefully, and two male *B. splendens* should not be housed in the same tank unless they are separated by a dividing wall. As a general rule, male bettas cannot be housed together. It is possible to house two male bettas in a single, very large tank, provided that there is plenty of cover (such as floating plants) and enough space for both males to establish their own territories. However, this is extremely risky because of the male's natural territoriality. Experiments in housing males together often end in the death of one or both inhabitants of the tank.

While they might eventually mate, keeping a male and female together may prove too volatile since the male will often be much more aggressive and mating conditions must be precisely conducive. Often, breeders have a special container so the female may display without being harmed by the male prior to induced breeding.

Females may or may not be able to coexist peacefully in the same tank depending on their temperaments. They are not schooling fish, and are still rather aggressive, but with enough room and many hiding spaces, they can learn to get along. There should never be exactly two female bettas in a tank together—a pecking order, a hierarchy, is necessary for them to live peacefully. With only two fish, one will be the bully and the other will be picked on. However, with three or more, a hierarchy is established.

Before co-housing Siamese fighting fish with other species, their compatibility should be carefully researched, and the owner should have a backup plan if the shared tank does not work. Common tankmates include mollies, catfish, or loaches. Amano Shrimp also provide good tankmates, and provided with sufficient natural plant cover will keep the tank clean without causing the betta any stress.

Although bettas are most aggressive towards each other, they have been known to kill very small fish or nip at the fins of fish such as fancy guppies, perhaps mistaking their finnage for that of another male betta. Certain fish should not be housed with bettas.

Schooling fish often become fin-nippers, making the betta a prime target because of their flowing fins. Also, aggressive fish like barbs should not be around bettas. Keepers have also reported problems when attempting to keep bettas in the company of piranha, or bluegill for obvious reasons. It is strongly recommended that bettas given tankmates should be housed in a tank that is at least 2 gallons per fish in the community (depending on bio load) with plenty of hiding places. Anything smaller will stress the betta. Only females can be kept in communities, and one still must watch out for aggressive females who will cause trouble in the tank.

Lifespan and Diet

Normally betta live to be 2-5 years old, but some live to be nearly 8 years old. However, when purchasing from a pet store, it is not always possible to know how old the fish is when you take it home. Typically, males are 9 months to one year old, as this is when their finnage becomes fullest and most attractive. Females can be much younger, as young as 3 to 6 months, due to their shorter finnage. Male bettas living in laboratories with large individual tanks and daily exercise have lived 10 years or longer.

Carnivorous, the betta feeds on zooplankton and mosquito and other insect larvae. Domesticated bettas will feed on bloodworms, daphnia, and brine shrimp.

Betta pellets are typically a combination of mashed shrimp meal, bloodworms, and various vitamins to enhance colour and longevity. For variety and fibre, bettas may also be fed finely chopped vegetables high in protein such as soybeans, green beans, broccoli, corn, or carrots.

Bettas are primarily surface feeders, because their mouths are upturned, so any food added should be able to float on the surface of the water.

Bettas fare better with a large variety of foods and will often show brighter, richer, and deeper colours if they are fed a wide range of foods. They will also heal much more quickly from fin damage if their diets are high in protein and fibre.

Two common maladies afflicting Siamese fighting fish are fin rot and ich.

Ldeal Tank Conditions

Since bettas are from the rice paddies of South East Asia, they typically thrive in conditions somewhat similar to their origins. In the wild, the Siamese fighting fish inhabits standing or slow-moving water, including floodplains and rice paddies, at temperatures of 24 to 30°C (75 to 86°F).

This level of temperature should be used in the aquarium. Colder water temperatures could lower the bettas' immune system and cause illness. The pH level should range between 6.5 and 7 (slightly acidic). Peat moss can be safely used to create softer acidic water since bettas are fine with water that is slightly tannic.

The most important factor in maintaining their ideal tank setup is that bettas require consistent conditions. They are easily stressed by sudden changes in temperature, *pH*, or bacteria. Also, bettas may be allergic to certain types of natural/ synthetic plastics, such as thermostats, elastomers, and thermoplastics.

Breeding

Bettas are fairly easy to breed if given healthy conditions. It is not unusual for one female and one male to produce more than a thousand viable eggs in one breeding; the male tends the eggs and newborns. However, after the fry are free swimming, the chore of tending the babies falls upon the human owner so preparation of baby food, baby-ready (cycled) tanks, and many other things must be considered and researched prior to actual breeding.

Silver Carp

The silver carp (*Hypophthalmichthys molitrix*) is a species of freshwater cyprinid fish, a variety of Asian carp native to north and northeast Asia. It is cultivated in China for food but was introduced to North America in the 1970s to control algae growth in aquaculture and municipal wastewater treatment facilities. They escaped from captivity soon after their importation. They are considered a highly invasive species and by 2003 had spread into the Mississippi, Illinois, Ohio and Missouri rivers and many of their tributaries. They are now (April 2007) abundant in the Mississippi River watershed from Louisiana to South Dakota and Illinois. Navigation dams on the Mississippi River seem to have slowed their advance up the Mississippi River, and few silver carp have been captured north of central Iowa on the Mississippi. Dams that do not have navigation locks are complete barriers to natural movement of silver carp, and it is important that fishermen do not assist this movement by the unintentional use of silver carp as bait.

The silver carp is also called the flying carp for its tendency to leap from the water when startled. They can grow to over 40 lb., and can leap 10 ft. in the air. Many boaters travelling in uncovered high-speed watercraft have been injured by running into the fish while at speed. In 2003, a woman jet-skiing broke her nose and a vertebra colliding with a silver carp and nearly drowned.

Pound for pound, more silver carp are produced worldwide in aquaculture than any other species. Silver carp are usually farmed in polyculture with other Asian major carps, or sometimes Indian major carps or other species. It has been introduced to, or spread into via connected waterways, at least 88 countries around the world. The most common reason for importation was for use in aquaculture, but enhancement of wild fisheries and water quality control were also important reasons for importation..

Diet

The silver carp is a filter feeder, and possesses a remarkably specialised filtration apparatus capable of filtering particles as small as 4 micrometres. The gill rakers are fused into a spongelike filter, and an epibranchial organ secretes mucous which assists in trapping small particles. A strong buccal pump forces water through this filter. Silver carp, like all *Hypophthalmichthys* species, have no stomachs; they are thought to feed more or less constantly. Silver carp are thought to feed largely on phytoplankton; they also consume zooplankton and detritus. Because of their plankton-feeding habits, there is concern that they will compete with native planktivorous fishes, which include paddlefish *Polyodon spathula*, gizzard shad *Dorosoma petenense* and young fish of almost all species.

Because they feed on plankton, they are sometimes successfully used as methods for controlling water quality, especially in the control of noxious bluegreen algae. However, these efforts are sometimes not successful. Certain species of bluegreen algae, notably the often toxic *Mycrocystis*, can pass through the gut of silver carp unharmed, and pick up nutrients while in the gut. Thus, in some cases bluegreen algae blooms have been exacerbated by silver carp. Also, *Mycrocystis* has been shown to produce more toxins in the presence of silver carp. Silver carp, which have natural defences to the toxins produced by bluegreen algae, sometimes can contain enough algal toxins in their systems that they become hazardous to eat.

Chanidae (about seven extinct species in five additional genera have been reported).

Milkfish have a generally symmetrical and streamlined appearance, with a sizable forked caudal fin. They can grow to 1.7 m, but are most often about 1 metre in length. They have no teeth, and generally feed on algae and invertebrates.

They occur in the Indian Ocean and across the Pacific Ocean, tending to school around coasts and islands with reefs. The youngest larvae live at sea for 2–3 weeks, then migrate to mangrove swamps, estuaries, and sometimes lakes, returning to sea to mature sexually and reproduce.

The larvae are collected from rivers and raised in ponds, where they can be fed almost anything and grow very quickly, then are sold either fresh, frozen, canned, or smoked.

The milkfish is also a national symbol of the Philippines, where it is called *bangus*. Because milkfish is notorious for being much more bony compared to other food fish in the country, deboned milkfish or "boneless bangus" has become popular and common in stores and markets.

Modoc Sucker

The Modoc sucker, *Catostomus microps*, is a species of fish native to California. It is listed as an endangered species in California and the United States, it is also listed as endangered by the IUCN.

Mojarra

The mojarras are a family, Gerreidae, of fishes in the order Perciformes.

Mojarras are a common prey and bait fish in many parts of the Caribbean including the South American Coast and Caribbean islands. These species tend to be difficult to identify in the field and often require microscopic examination. Most species exhibit a schooling behaviour and tend to exploit the shallow water refugia associated with coastal areas presumably to avoid large-bodied predators.

Genera

- *Diapterus* (Ranzani, 1842)
- *Eucinostomus* (Baird and Girard in Baird, 1855)
- *Eugerres* (Jordan and Evermann, 1927)
- *Gerres* (Quoy and Gaimard, 1824)
- *Parequula* (Steindachner, 1879)
- *Pentaprion* (Bleeker, 1850)
- *Ulaema* (Jordan and Evermann in Jordan, 1895)

Poecilia

Poecilia is a genus of freshwater fish in family Poeciliidae of order Cyprinodontiformes. The type species is *P. vivipara*. Live-bearers, the *Poecilia* species are collectively known as mollies, with the exception of Endler's livebearer (*P. wingei*) and the famous guppy (*P. reticulata*). Members of this genus are members of the family Poecilidae, which includes the southern platyfish or "platy" (*Xiphophorus maculatus*), and the green swordtail (*X. hellerii*).

IUCN list two of the species, the sulphur molly, *P. sulphuraria*, and the broadspotted molly, *P. latipunctata*, as Critically Endangered.

Species

FishBase lists 33 species. Recently one new species was added, bringing the total to 34 species:

- *Poecilia amazonica* (Garman, 1895).
- *Poecilia boesemani* (Poeser, 2003).
- Pacific molly, *Poecilia butleri* (Jordan, 1889).
- Catemaco molly, *Poecilia catemaconis* (Miller, 1975).
- *Poecilia caucana* (Steindachner, 1880).
- *Poecilia caudofasciata* (Regan, 1913).
- Dwarf molly, *Poecilia chica* (Miller, 1975).
- *Poecilia dauli* (Meyer & Radda, 2000).
- Elegant molly, *Poecilia elegans* (Trewavas, 1948).

- Amazon molly, *Poecilia formosa* (Girard, 1859).
- *Poecilia gillii* (Kner, 1863).
- Hispaniola molly, *Poecilia hispaniolana* (Rivas, 1978).
- *Poecilia koperi* (Poeser, 2003).
- *Poecilia kykesis* (Poeser, 2002).
- Sailfin molly, *Poecilia latipinna* (Lesueur, 1821).
- Broadspotted molly, *Poecilia latipunctata* (Meek, 1904).
- *Poecilia marcellinoi* (Poeser, 1995).
- Balsas molly, *Poecilia maylandi* (Meyer, 1983).
- *Poecilia mechthildae* (Bork, Etzel & Meyer, 2002).
- Shortfin molly, *Poecilia mexicana* (Steindachner, 1863).
- *Poecilia nicholsi* (Myers, 1931).
- Mangrove molly, *Poecilia orri* (Fowler, 1943).
- Peten molly, *Poecilia petenensis* (Gunther, 1866).
- Guppy, *Poecilia reticulata* (Peters, 1859).
- *Poecilia rositae* (Meyer, Radda, Schartl, Schneider & Wilde, 2004).
- *Poecilia salvatoris* (Regan, 1907).
- Molly, black molly *Poecilia sphenops* (Valenciennes , 1846).
- Sulphur molly, *Poecilia sulphuraria* (Alvarez, 1948).
- Mountain molly, *Poecilia teresae* (Greenfield, 1990).
- *Poecilia vandepolli* (Van Lidth de Jeude, 1887).
- Sail-fin molly, *Poecilia velifera* (Regan, 1914).
- *Poecilia vivipara* (Bloch & Schneider, 1801).
- *Poecilia wandae* (Poeser, 2003).
- Endler's livebearer, *Poecilia wingei* (Isbrucker, Kempkes & Poeser, 2005).

Monkfish

Monkfish is a common English name of a number of species of fish.

Lophius

Most of the fish referred to as monkfish belong to the genus *Lophius*, in the anglerfish family Lophiidae. Monkfish is the most common English name for this genus in the northwest Atlantic but goosefish is used as the equivalent term on the eastern coast of North America. As the most common anglerfish found in coastal waters around the British Isles, it can be known by that name alone.

This monkfish has three long filaments sprouting from the middle of its head; these are the detached and modified three first spines of the anterior dorsal fin. As with all anglerfish species, the longest filament is the first, which terminates in an irregular growth of flesh, and is movable in all directions; this "tentacle" is used as a lure to attract other fishes, which the monkfish then seizes with its enormous jaws, devouring them whole.

Experiments have shown, however, that whether the prey has been attracted to the lure or not is not strictly relevant, as the action of the jaws is an automatic reflex triggered by contact with the tentacle.

It grows to a length of more than 5 ft.; specimens of 3 ft. are common.

Two species *Lophius piscatorius* and *Lophius budegessa* are found in north-western Europe, and referred to as monkfish. With *L. piscatorius* by far the most common species around the British Isles.

Squatinidae

A second group of fish also known as monkfish are members of the genus *Squatina*, in the angel shark family Squatinidae. These are of somewhat similar shape to the anglerfish, but completely unrelated; like the true sharks, they are elasmobranchs. These fish are only of minor significance for human consumption, though they are endangered because they are caught as bycatch by trawlers.

A Supposed Monk-Like Sea Monster: The term monkfish has also been used for a sea monster of the north-west Atlantic bearing a passing resemblance to a monk (also known as a sea monk).

Mooneye

The mooneyes are a family Hiodontidae of primitive ray-finned fish comprising two living and one extinct species in the genus *Hiodon*. They are large-eyed, fork-tailed fish that physically resemble shads somewhat. Their common name comes from the metallic gold or silver shine of their eyes.

The mooneye, *Hiodon tergisus* Lesueur 1818, is widespread across North America, living in the clear waters of lakes, ponds, and rivers. It consumes aquatic invertebrates, insects, and fish. Mooneyes can reach 47 cm in length.

Mooneyes feed readily on terrestrial insects, and will provide fine sport for an intrepid flyrod angler. They will also take small lures and natural baits with gusto. Mooneyes are frantic, hyperactive fish and their impressive leaps and passionate fighting style has earned them the nickname "Freshwater Tarpon".

The goldeye, *Hiodon alosoides* Rafinesque 1819, is also widespread across North America, and is notable for a conspicuous golden iris in the eyes. It prefers turbid slower-moving waters of lakes and rivers, where it feeds on insects, crustaceans, fish, frogs, shrews, and mice. The fish has been reported up to 52 cm in length. The smoked meat is highly valued and sold as "Winnipeg goldeye".

Moorish Idol

The moorish idol, *Zanclus cornutus* ("Crowned Scythe"), is a small perciform marine fish, the sole representative of the family Zanclidae (from the Greek *zagkios*, "oblique"). A common inhabitant of tropical to subtropical reefs and lagoons, the moorish idol is notable for its wide distribution throughout the Indo-Pacific. A number of butterflyfishes (all of the genus *Heniochus*) closely resemble the moorish idol.

It is said the moorish idol got its name from the Moors of Africa, who purportedly believe the fish to be a bringer of happiness. Moorish idols are also popular aquarium fish, but despite their popularity, they are notorious for their short aquarium lifespans and difficulty.

Physical Description

With distinctively compressed and disk-like bodies, moorish idols stand out in contrasting bands of black, white and yellow which make them look very attractive to aquarium keepers. The fish have relatively small fins, except for the dorsal fin whose 6 or 7 spines are dramatically elongated to form a trailing, sickle-shaped crest called the philomantis extension. Moorish idols have small terminal mouths at the end of long, tubular snouts; many long bristle-like teeth line the mouth.

The eyes are set high on the fish's deeply-keeled bodies; in adults, perceptible bumps are located above each. The anal fin may have 2 or 3 spines. Moorish idols reach a maximum length of 23 cm. The sickle-like dorsal spines actually shorten with age.

Habitat and Diet

Generally denizens of shallow waters, moorish idols prefer flat reefs. The fish may be found at depths from 3 to 180 m, in both murky and clear conditions. The range of the moorish idol includes East Africa and the Ducie Islands; Hawaii, southern Japan and all of Micronesia; they are also found from the southern Gulf of California south to Peru.

Sponges, tunicates and other benthic invertebrates constitute the bulk of the moorish idol's diet. Captive kept moorish idols typically are very picky eaters. They will either eat nothing (common) and perish or eat everything (very uncommon). Eating a variety of items is healthy. Even small portions of avocado and banana are sometimes fed in captivity.

Behaviour and Reproduction

Often glimpsed alone, moorish idols will also form pairs or occasionally small schools. They are diurnal fish, sticking to the bottom of the reef at night and adopting a drab colouration. Like the butterflyfishes, moorish idols mate for life; as juveniles, they are more apt to school. Adult males tend to be aggressive towards one another.

Moorish idols are pelagic spawners; that is, eggs and sperm are released in midwater and the fertilized eggs are left to drift away with the currents. The impressive range of these fish may be explained by the unusually long larval stage; the fish reach a length of 7.5 cm before becoming free-swimming juveniles. Before this time, the developing larvae will have drifted considerable distances.

Aquarium Life

Moorish idols are notorious for being difficult to maintain in captivity. They require enormous tanks, often exceeding 200 US gal, are voracious eaters, and are infamous for becoming incredibly destructive. Their captive survival rate is very low: most do not survive for a full year. Most that live past this mark typically die shortly thereafter. It is not recommended that any aquarist attempt to keep this species, because it is considered cruel by many and is nearly impossible. To avoid these shortfalls, some aquarists prefer to keep substitute species that look very similar to the Moorish Idol. These substitutes are all butterflyfishes of the genus *Heniochus*, and include the pennant coralfish, *Heniochus acuminatus*; threeband pennantfish, *H. chrysostomus*; and the false moorish idol, *H. diphreutes*.

Moray eel

Moray eels are large cosmopolitan eels of the family Muraenidae. There are approximately 200 species in 15 genera. The typical length for a moray is 1.5m (5 feet), with the largest being the slender giant moray, *Strophidon sathete*, at up to 4 m (13 feet).

Moray eels frequent tropical and subtropical coral reefs to depths of 200 m, where they spend most of their time concealed inside crevices and alcoves. They secrete a protective mucus over their scaleless skin which contains a toxin in some species.

Their small circular gills, located on the flanks far posterior to the mouth, require the moray eel to maintain a gape in order to facilitate respiration.

The dorsal fin of the moray extends from just behind the head, along the back and joins seamlessly with the caudal and anal fin. Most species lack pectoral and pelvic fins, adding to their snake-like appearance. Their eyes are rather small; morays rely on their highly developed sense of smell, lying in wait to ambush prey.

The body of the Moray Eel is patterned, camouflage also being present inside the mouth. Their jaws are wide, with a snout that protrudes forward. They possess large teeth, designed to tear flesh as opposed to holding or chewing. They are capable of inflicting serious wounds to humans.

Morays are carnivorous and feed primarily on other fish, cephalopods, molluscs, and crustaceans. Groupers, other moray eels, and barracudas are among their few predators. There is a commercial fishery for several species, but some have been known to cause ciguatera fish poisoning. Morays hide in crevices in the reefs, and wait until their prey is close enough for capture. They then jump out and clamp the prey in their strong jaws.

Cooperative Hunting

In the December 2006 issue of the journal, *Public Library of Science Biology*, a team of biologists announced the discovery of interspecies cooperative hunting involving Morays. The biologists, who were engaged in a study of Red Sea cleaner-fish (fish that enter the mouths of other fish to rid them of parasites), discovered that a species of grouper, *Plectropomus Pessuliferus*, often recruited moray eels to aid them while hunting for food.

This is the first discovery of cooperation between fish in general, and the first known inter-species cooperation outside of humans and dogs, humans and falcons, humans and cats, and humans and dolphins.

Genera

- *Anarchias*
- *Channomuraena*
- *Cirrimaxilla*
- *Echidna*
- *Enchelycore*
- *Enchelynassa*
- *Gymnomuraena*
- *Gymnothorax*
- *Monopenchelys*
- *Muraena*
- *Pseudechidna*
- *Rhinomuraena*
- *Scuticaria*
- *Strophidon*
- *Uropterygius*

Moridae

Moridae is a family of cod-like fishes, known as codlings, hakelings and moras. They are marine fishes found throughout the world, and grow up to 90 cm long (red codling, *Pseudophycis bachus*).

Species

There are 111 species in 18 genera:

- Genus *Antimora*
  - — Finescale mora, *Antimora microlepis* (Bean, 1890).

Finescale mora, Antimora microlepis

— Blue antimora or Violet cod, *Antimora rostrata* (Gunther, 1878).

- Genus *Auchenoceros*

— Ahuru, *Auchenoceros punctatus* (Hutton, 1873).

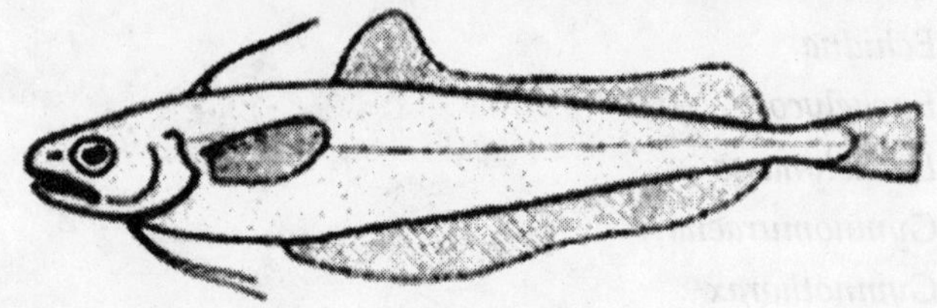

Ahuru, Auchenoceros punctatus

- Genus *Eeyorius*

— Tasmanian codling, *Eeyorius hutchinsi* (Paulin, 1986).

- Genus *Eretmophorus*

— *Eretmophorus kleinenbergi* (Giglioli, 1889).

- Genus *Gadella*

— *Gadella brocca* (Paulin & Roberts, 1997).

— *Gadella dancoheni* (Sazonov & Shcherbachev, 2000).

— *Gadella edelmanni* (Brauer, 1906).

— *Gadella filifer* (Garman, 1899).

— Beardless codling, *Gadella imberbis* (Vaillant, 1888).

— *Gadella jordani* (Bohlke & Mead, 1951).

— *Gadella macrura* (Sazonov & Shcherbachev, 2000).

— Gadella, *Gadella maraldi* (Risso, 1810).

Gadella, Gadella Maraldi

— *Gadella molokaiensis* (Paulin, 1989).

— *Gadella norops* (Paulin, 1987).
— *Gadella obscurus* (Parin, 1984).
— *Gadella svetovidovi* (Trunov, 1992).
— *Gadella thysthlon* (Long & McCosker, 1998).

• Genus *Guttigadus*

— Fat-headed cod, *Guttigadus globiceps* (Gilchrist, 1906).
— Tadpole cod, *Guttigadus globosus* (Paulin, 1986).
— *Guttigadus kongi* (Markle & Melendez C., 1988).
— *Guttigadus latifrons* (Holt & Byrne, 1908).
— *Guttigadus nudicephalus* (Trunov, 1990).
— *Guttigadus nudirostre* (Trunov, 1990).
— *Guttigadus squamirostre* (Trunov, 1990).

• Genus *Halargyreus*

— Slender codling or Slender cod, *Halargyreus johnsonii* (Gunther, 1862).

Slender codling, Halargyreus johnsonii

• Genus *Laemonema*

— Shortbeard codling, *Laemonema barbatulum* (Goode & Bean, 1883).

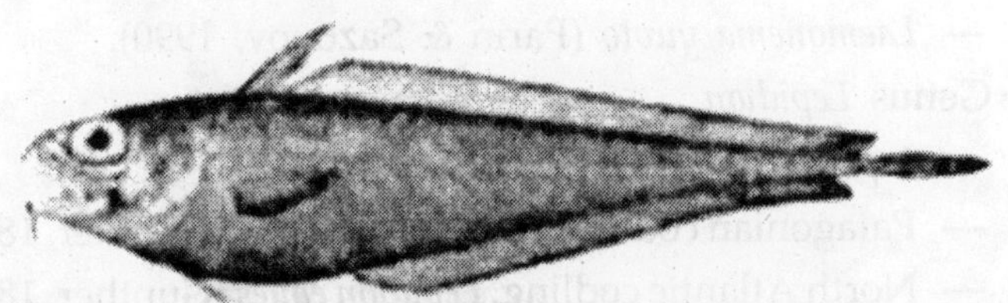

Shortbeard codling, Laemonema barbatulum

— *Laemonema compressicauda* (Gilchrist, 1903).

— *Laemonema filodorsale* (Okamura, 1982).
— *Laemonema goodebeanorum* (Melendez C. & Markle, 1997).
— *Laemonema gracillipes* (Garman, 1899).
— Guinean codling, *Laemonema laureysi* (Poll, 1953).
— Longfin codling, *Laemonema longipes* (Schmidt, 1938).
— *Laemonema macronema* (Melendez C. & Markle, 1997).
— *Laemonema melanurum* (Goode & Bean, 1896).

Laemonema melanurum

— *Laemonema modestum* (Franz, 1910).
— *Laemonema nana* (Taki, 1953).
— *Laemonema palauense* (Okamura, 1982).
— *Laemonema rhodochir* (Gilbert, 1905).
— Robust mora, *Laemonema robustum* (Johnson, 1862).
— Bighead mora, *Laemonema verecundum* (Jordan & Cramer, 1897).
— *Laemonema yarrellii* (Lowe, 1838).
— *Laemonema yuvto* (Parin & Sazonov, 1990).

- Genus *Lepidion*

— *Lepidion capensis* (Gilchrist, 1922).
— Patagonian codling, *Lepidion ensiferus* (Gunther, 1887).
— North Atlantic codling, *Lepidion eques* (Gunther, 1887).
— *Lepidion guentheri* (Giglioli, 1880).
— Morid cod, *Lepidion inosimae* (Gunther, 1887).
— Mediterranean codling, *Lepidion lepidion* (Risso, 1810).

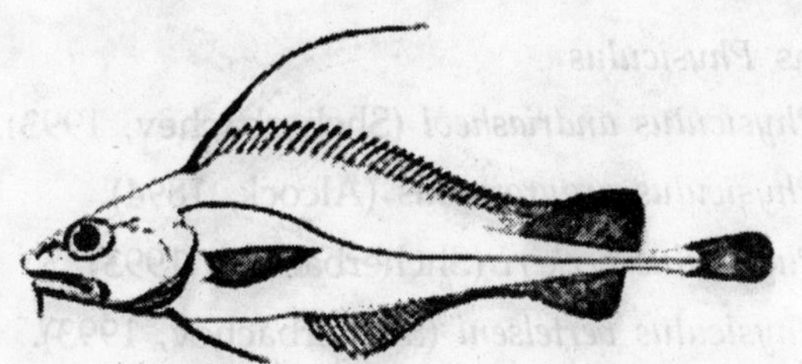

Mediterranean codling, Lepidion lepidion

— Small-headed cod or Long-finned cod, *Lepidion microcephalus* (Cowper, 1956).

— *Lepidion natalensis* (Gilchrist, 1922).

— *Lepidion schmidti* (Svetovidov, 1936).

• Genus *Lotella*

— *Lotella fernandeziana* (Rendahl, 1921).

— *Lotella fuliginosa* (Gunther, 1862).

— Beardie, *Lotella phycis* (Temminck & Schlegel, 1846).

— Rock cod, *Lotella rhacina* (Forster, 1801).

— Slender beardy, *Lotella schuettei* (Steindachner, 1866).

— *Lotella tosaensis* (Kamohara, 1936).

• Genus *Mora*

— Common mora, googly-eyed cod or ribaldo, *Mora moro* (Risso, 1810).

Common mora, Mora moro

• Genus *Notophycis*

— *Notophycis fitchi* (Sazonov, 2001).

— Dwarf codling, *Notophycis marginata* (Gunther, 1878).

- Genus *Physiculus*
  - — *Physiculus andriashevi* (Shcherbachev, 1993).
  - — *Physiculus argyropastus* (Alcock, 1894).
  - — *Physiculus beckeri* (Shcherbachev, 1993).
  - — *Physiculus bertelseni* (Shcherbachev, 1993).
  - — *Physiculus capensis* (Gilchrist, 1922).
  - — *Physiculus chigodarana* (Paulin, 1989).
  - — *Physiculus coheni* (Paulin, 1989).
  - — *Physiculus cyanostrophus* (Anderson & Tweddle, 2002).
  - — *Physiculus cynodon* (Sazonov, 1987).
  - — Black codling, *Physiculus dalwigki* (Kaup, 1858).
  - — *Physiculus fedorovi* (Shcherbachev, 1993).
  - — Hakeling, *Physiculus fulvus* (Bean, 1884).

Hakeling, Physiculus fulvus

  - — *Physiculus grinnelli* (Jordan & Jordan, 1922).
  - — Skulpin, *Physiculus helenaensis* (Paulin, 1989).
  - — *Physiculus hexacytus* (Parin, 1984).
  - — *Physiculus huloti* (Poll, 1953).
  - — Japanese codling, *Physiculus japonicus* (Hilgendorf, 1879).
  - — *Physiculus karrerae* (Paulin, 1989).
  - — *Physiculus kaupi* (Poey, 1865).

Physiculus kaupi

— *Physiculus longicavis* (Parin, 1984).
— *Physiculus longifilis* (Weber, 1913).
— Luminescent cod, *Physiculus luminosus* (Paulin, 1983).
— *Physiculus marisrubri* (Bruss, 1986).
— *Physiculus maslowskii* (Trunov, 1991).
— *Physiculus microbarbata* (Paulin & Matallanas, 1990).
— *Physiculus natalensis* (Gilchrist, 1922).
— Charcoal mora, *Physiculus nematopus* (Gilbert, 1890).
— *Physiculus nielseni* (Shcherbachev, 1993).
— *Physiculus nigripinnis* (Okamura, 1982).
— *Physiculus nigriscens* (Smith & Radcliffe, 1912).
— *Physiculus normani* (Bruss, 1986).
— *Physiculus parini* (Paulin, 1991).
— *Physiculus peregrinus* (Gunther, 1872).
— Hundred fathom mora, *Physiculus rastrelliger* (Gilbert, 1890).
— *Physiculus rhodopinnis* (Okamura, 1982).
— *Physiculus roseus* (Alcock, 1891).
— *Physiculus sazonovi* (Paulin, 1991).
— *Physiculus sterops* (Paulin, 1989).
— *Physiculus sudanensis* (Paulin, 1989).
— Peruvian mora, *Physiculus talarae* (Hildebrand & Barton, 1949).
— *Physiculus therosideros* (Paulin, 1987).
— *Physiculus yoshidae* (Okamura, 1982).

- Genus *Pseudophycis*
  - — Red codling, *Pseudophycis bachus* (Forster, 1801).
  - — Southern bastard codling, *Pseudophycis barbata* (Gunther, 1863).
  - — Northern bastard codling, *Pseudophycis breviuscula* (Richardson, 1846).
- Genus *Rhynchogadus*
  - — *Rhynchogadus hepaticus* (Facciola, 1884).
- Genus *Salilota*
  - — Tadpole codling, *Salilota australis* (Barnard, 1925).
- Genus *Svetovidovia*
  - — *Svetovidovia lucullus* (Jensen, 1953).
- Genus *Tripterophycis*
  - — Grenadier cod, *Tripterophycis gilchristi* (Boulenger, 1902).

Morwong

Morwongs are perciform fishes comprising the family Cheilodactylidae. Found primarily throughout the Southern Hemisphere, they are also found in the Pacific off Japan, China, and Hawaii. Growing up to 1 metre in length, they feed on small invertebrates on the ocean floor. Several species of morwong are commercially harvested as food fish, particularly in Australia.

Other names for members of the family include butterfish, fingerfin, jackassfish, and moki. Morwongs are also erroneously known as snappers.

Morwong is also used as a name for several unrelated fish found in Australian waters, such as the painted sweetlips, *Diagramma pictum*.

Genera and Species

- Genus *Cheilodactylus*
  - — Redfingers, *Cheilodactylus fasciatus*

The smalltooth sawfish has a long, flat, blade-like rostrum with 24 to 32 pairs of teeth along the edges. The caudal fin is large and oblique with no lower lobe. It is an inshore and intertidal species, but may cross deep water to reach offshore islands, but also ascends rivers and can tolerate fresh water. It is commonly seen in bays, lagoons, estuaries, and river mouths, but is also found in rivers and lakes. It uses its saw to stir the bottom when feeding on bottom invertebrates and to kill pelagic fishes.

It is utilised as a food fish, the oil being used to make medicine, soap and in leather tanning. Adults are stuffed for decoration. It is reported to be aggressive towards sharks when kept in tanks.

Colouration is dark mouse grey to blackish brown above, paler along margins of fins, and white to greyish white or pale yellow below.

Smelts

Smelts are a family, Osmeridae, of small anadromous fish. They are common in the North American Great Lakes, and run in large schools along the coastline during their spring migration to their spawning streams. The family consists of some sixteen species in six genera.

The fish usually reach only 6 inches (15 cm) and are a food source for salmon and lake trout. It is one of the few fish that sportsmen are allowed to net, using dip nets, either along the coastline or in the streams. Some sportsmen also ice fish for smelt. Smelt are often fried and eaten whole.

Smelt roe is bright orange in colour, and is often used to garnish sushi.

In Michigan and other northern states, "smelt dipping" is a common sport in the early spring months (generally late April in the Upper Peninsula, when the stream water reaches appx. 40 degrees F). Fish are simply spotted using a flashlight (the best smelt dipping is in the middle of the night) and scooped out of the water using a dip net made of nylon or metal mesh. To clean a smelt, simply remove the head and the entrails. Fins, scales, and bones of all but the largest of smelts will soften when the smelts are cooked and do not need to be removed.

On the Maine coast, smelts were also a sign of spring, with the run of these small fish up tiny tidal estuaries. Many of these 'rivers' were small enough that a person could straddle the water and, leaning over, dip a bucket and get a good catch of smelt. This was a night-time operation, and people might line up to get their time over the stream.

Miscellaneous

Smelts were traditionally an important winter catch in the salt water mouths of rivers in New England and the Maritime Provinces of Canada. Fishermen would go to customary locations over the ice using horses and sleighs. Smelt taken out of the cold salt water were much preferred to those taken in warm water. The smelt did not command a high price on the market, but provided a useful supplemental income in times when wants were much less. The smelts were "flash frozen" simply by leaving them on the ice and then sold to fish buyers who came down the rivers on horse and sleigh. They were also an excellent winter meal. They were gutted, heads and tails removed and rinsed in cold water then dipped in flour mixed with salt and pepper and fried in butter. Served with boiled potatoes and pickled beets, they were a welcome addition to winter fare.

A variety of Smelt, the Delta Smelt is found in the Sacramento-San Joaquin Delta. The Delta Smelt is a protected species. Catholics of Italian heritage on the Northeastern coast of America often eat smelts as part of Christmas Eve dinner.

Smooth Dogfish

The smooth dogfish, *Mustelus canis*, is a species of shark. This shark is an olive grey or brown, and may have shades of yellow or greyish white. Females live to 16 years and males have a life span of 10 years.

Size and Growth

Length for the smooth dogfish is up to 1.5 m (60 inches), with a maximum weight of 12 kg (27 lb). Smooth dogfish reach maximum size at 7 or 8 years of age. Average size of this shark is approximately 1.2 m (48 in). This species grows quickly, with males reaching maturity at 2 or 3 years of age, and females at 4 to five years of age.

Habitat

A common resident in bays, and other inshore waters, the smooth dogfish prefers shallow waters of less than 18 m (60 ft) in depth but may be found to depths of 200 m (650 ft). This species has also been found on occasion in freshwater although it is unlikely they can survive freshwater for extended periods of time. The smooth dogfish migrates seasonally, moving north in the spring and south in the autumn. It is primarily a nocturnal species.

Food

A scavenger and opportunistic predator, the smooth dogfish feeds primarily on large crustaceans, including lobsters, shrimp, and crabs, as well as small fish and molluscs. The flat, blunt teeth of the dogfish are used to crush and grind these prey items which have tough outer body coverings. Small fish that are preyed upon by the smooth dogfish include menhaden and tautog. Young smooth dogfish feed on small shrimps, worms, and crabs.

Reproduction

Mating occurs throughout most of the smooth dogfish's range from May through July. Following a gestation period of approximately 10 or 11 months, a litter numbering as few as 4 and as many as 20 is born during late spring or early summer. Larger females tend to have larger litters.

Importance to Humans

In certain areas, the flesh of smooth dogfish is marketed as fresh or dried salted for human consumption. The smooth dogfish is often used as a laboratory animal and in public display at aquariums.

Snake Mackerel

Snake mackerel are perciform fishes in the family Gempylidae. They are elongate fishes with a similar appearance to barracudas, having a long dorsal fin, usually with one or finlets trailing it. The largest species, including the snoek, *Thyrsites atun*, grow up to two metres long.

They are deep-water benthopelagic fishes, and several species are important commercial and game fishes.

Species

There are 24 species in sixteen genera:

- Genus *Diplospinus*
  - — Striped escolar, *Diplospinus multistriatus* (Maul, 1948).
- Genus *Epinnula*
  - — Domine, *Epinnula magistralis* (Poey, 1854).
- Genus *Gempylus*
  - — *Gempylus serpens* (Cuvier, 1829).
- Genus *Lepidocybium*
  - — Escolar, *Lepidocybium flavobrunneum* (Smith, 1843).
- Genus *Nealotus*
  - — Black snake mackerel, *Nealotus tripes* (Johnson, 1865).
- Genus *Neoepinnula*
  - — American sackfish, *Neoepinnula americana* (Grey, 1953).
  - — Sackfish, *Neoepinnula orientalis* (Gilchrist & von Bonde, 1924).
- Genus *Nesiarchus*
  - — Black gemfish, *Nesiarchus nasutus* (Johnson, 1862).
- Genus *Paradiplospinus*
  - — Antarctic escolar, *Paradiplospinus antarcticus* (Andriashev, 1960).
  - — Slender escolar, *Paradiplospinus gracilis* (Brauer, 1906).

- Genus *Promethichthys*
- Genus *Rexea*
  — *Rexea alisae* (Roberts & Stewart, 1997).
  — Long-finned escolar, *Rexea antefurcata* (Parin, 1989).
  — Bengal escolar, *Rexea bengalensis* (Alcock, 1894).
  — Short-lined escolar, *Rexea brevilineata* (Parin, 1989).
  — Nakamura's escolar, *Rexea nakamurai* (Parin, 1989).
  — Royal escolar, *Rexea prometheoides* (Bleeker, 1856).
  — Silver gemfish or Gemfish, *Rexea solandri* (Cuvier, 1832).
- Genus *Rexichthys*
  — Paxton's escolar, *Rexichthys johnpaxtoni* (Parin & Astakhov, 1987).
- Genus *Ruvettus*
  — Oilfish, *Ruvettus pretiosus* (Cocco, 1833).
- Genus *Thyrsites*
  — Snoek, *Thyrsites atun* (Euphrasen, 1791).
- Genus *Thyrsitoides*
  — Black snoek, *Thyrsitoides marleyi* (Fowler, 1929).
- Genus *Thyrsitops*
  — White snake mackerel, *Thyrsitops lepidopoides* (Cuvier, 1832).
- Genus *Tongaichthys*
  — Tonga escolar, *Tongaichthys robustus* (Nakamura & Fujii, 1983).

The longfin escolar, *Scombrolabrax heterolepis*, was formerly classified in this family, but it is now placed in its own family Scombrolabracidae.

Snapper

The snapper are a family of perciform fishes, mainly marine but with some members living in estuaries, and entering fresh water to feed. Some are important food fish. One of the best known is the red snapper.

Snappers are found in the tropical and subtropical regions of all the oceans. They can grow to about a metre in length. Most feed on crustaceans or other fish, though a few are plankton-feeders. They can be kept in aquaria, but mostly grow too fast to be popular aquarium fish. They live at depths of up to 450 m.

About 100 species of snapper are currently recognised, divided into about 16 genera. A very large number of fish species have "snapper" in their common name; most but not all of these are members of the family Lutjanidae. Almost all the 60 or so species in the genus *Lutjanus* have common names including the word "snapper".

Unrelated to the fishes, the word "snapper" is also used for the snapping turtles, as for the common snapper and the alligator snapper.

Species

- Genus *Aphareus*
  - — Small toothed jobfish, *Aphareus furca* (Lacepede, 1801).
  - — Rusty jobfish, *Aphareus rutilans* (Cuvier, 1830).
- Genus *Aprion*
  - — Green jobfish, *Aprion virescens* (Valenciennes, 1830).
- Genus *Apsilus*
  - — Black snapper, *Apsilus dentatus* (Guichenot, 1853).
  - — African forktail snapper, *Apsilus fuscus* (Valenciennes, 1830).
- Genus *Etelis*
  - — Ruby snapper, *Etelis carbunculus* (Cuvier, 1828).
  - — Flame snapper, *Etelis coruscans* (Valenciennes, 1862).
  - — Queen snapper, *Etelis oculatus* (Valenciennes, 1828).
  - — Scarlet snapper, *Etelis radiosus* (Anderson, 1981).
- Genus *Hoplopagrus*
  - — Mexican barred snapper, *Hoplopagrus guentherii* (Gill, 1862).

- Genus *Lipocheilus*
  - — Tang's snapper, *Lipocheilus carnolabrum* (Chan, 1970).
- Genus *Lutjanus*
  - — Yellow-banded snapper, *Lutjanus adetii* (Castelnau, 1873).
  - — African red snapper, *Lutjanus agennes* (Bleeker, 1863).
  - — *Lutjanus alexandrei* (Moura & Lindeman, 2007).
  - — Ambiguous snapper, *Lutjanus ambiguus* (Poey, 1860).
  - — Mutton snapper, *Lutjanus analis* (Cuvier, 1828).
  - — Schoolmaster snapper, *Lutjanus apodus* (Walbaum, 1792).
  - — Mullet snapper, *Lutjanus aratus* (Gunther, 1864).
  - — Mangrove red snapper, *Lutjanus argentimaculatus* (Forsskal, 1775).
  - — Yellow snapper, *Lutjanus argentiventris* (Peters, 1869).
  - — Bengal snapper, *Lutjanus bengalensis* (Bloch, 1790).
  - — Two-spot banded snapper, *Lutjanus biguttatus* (Valenciennes, 1830).
  - — Indonesian snapper, *Lutjanus bitaeniatus* (Valenciennes, 1830).
  - — Two-spot red snapper, *Lutjanus bohar* (Forsskal, 1775).
  - — Moluccan snapper, *Lutjanus boutton* (Lacepede, 1802).
  - — Blackfin snapper, *Lutjanus buccanella* (Cuvier, 1828).
  - — Northern red snapper, *Lutjanus campechanus* (Poey, 1860).
  - — Spanish flag snapper, *Lutjanus carponotatus* (Richardson, 1842).
  - — Blueline snapper, *Lutjanus coeruleolineatus* (Ruppell, 1838).
  - — Colorado snapper, *Lutjanus colorado* (Jordan & Gilbert, 1882).
  - — Cubera snapper, *Lutjanus cyanopterus* (Cuvier, 1828).
  - — Checkered snapper, *Lutjanus decussatus* (Cuvier, 1828).

— African brown snapper, *Lutjanus dentatus* (Dumeril, 1861).

— Sunbeam snapper, *Lutjanus dodecacanthoides* (Bleeker, 1854).

— Blackspot snapper, *Lutjanus ehrenbergii* (Peters, 1869).

— Guinea snapper, *Lutjanus endecacanthus* (Bleeker, 1863).

— Crimson snapper, *Lutjanus erythropterus* (Bloch, 1790).

— Golden African snapper, *Lutjanus fulgens* (Valenciennes, 1830).

— Dory snapper, *Lutjanus fulviflamma* (Forsskal, 1775).

— Blacktail snapper, *Lutjanus fulvus* (Forster, 1801).

— Freshwater snapper, *Lutjanus fuscescens* (Valenciennes, 1830).

— Humpback red snapper, *Lutjanus gibbus* (Forsskal, 1775).

— Papuan black snapper, *Lutjanus goldiei* (Macleay, 1882).

— Gorean snapper, *Lutjanus goreensis* (Valenciennes, 1830).

— Mangrove snapper, *Lutjanus griseus* (Linnaeus, 1758).

— Yellowfin red snapper, *Lutjanus guilcheri* (Fourmanoir, 1959).

— Spotted rose snapper, *Lutjanus guttatus* (Steindachner, 1869).

— Golden snapper, *Lutjanus inermis* (Peters, 1869).

— Dog snapper, *Lutjanus jocu* (Bloch & Schneider, 1801).

— John's snapper, *Lutjanus johnii* (Bloch, 1792).

— Jordan's snapper, *Lutjanus jordani* (Gilbert, 1898).

— Common bluestripe snapper, *Lutjanus kasmira* (Forsskal, 1775).

— Yellowstreaked snapper, *Lutjanus lemniscatus* (Valenciennes, 1828).

— Lunartail snapper, *Lutjanus lunulatus* (Park, 1797).

— Bigeye snapper, *Lutjanus lutjanus* (Bloch, 1790).

— Indian snapper, *Lutjanus madras* (Valenciennes, 1831).

— Mahogany snapper, *Lutjanus mahogoni* (Cuvier, 1828).

— Malabar blood snapper, *Lutjanus malabaricus* (Bloch & Schneider, 1801).

— Pygmy snapper, *Lutjanus maxweberi* (Popta, 1921).

— Samoan snapper, *Lutjanus mizenkoi* (Allen & Talbot, 1985).

— Onespot snapper, *Lutjanus monostigma* (Cuvier, 1828).

— Bluestriped snapper, *Lutjanus notatus* (Cuvier, 1828).

— Pacific cubera snapper, *Lutjanus novemfasciatus* (Gill, 1862).

— Spotstripe snapper, *Lutjanus ophuysenii* (Bleeker, 1860).

— Pacific red snapper, *Lutjanus peru* (Nichols & Murphy, 1922).

— Southern red snapper, *Lutjanus purpureus* (Poey, 1876).

— Five-lined snapper, *Lutjanus quinquelineatus* (Bloch, 1790).

— Blubberlip snapper, *Lutjanus rivulatus* (Cuvier, 1828).

— Yellow-lined snapper, *Lutjanus rufolineatus* (Valenciennes, 1830).

— Russell's snapper, *Lutjanus russellii* (Bleeker, 1849).

— Humphead snapper, *Lutjanus sanguineus* (Cuvier, 1828).

— Emperor red snapper, *Lutjanus sebae* (Cuvier, 1816).

— Black-banded snapper, *Lutjanus semicinctus* (Quoy & Gaimard, 1824).

— Star snapper, *Lutjanus stellatus* (Akazaki, 1983).

— Lane snapper, *Lutjanus synagris* (Linnaeus, 1758).

— Timor snapper, *Lutjanus timorensis* (Quoy & Gaimard, 1824).

— Blue and gold snapper, *Lutjanus viridis* (Valenciennes, 1846).

— Brownstripe red snapper, *Lutjanus vitta* (Quoy & Gaimard, 1824).

— Silk snapper, *Lutjanus vivanus* (Cuvier, 1828).

- Genus *Macolor*

— Midnight snapper, *Macolor macularis* (Fowler, 1931).

— Black and white snapper, *Macolor niger* (Forsskal, 1775).

- Genus *Ocyurus*

— Yellowtail snapper, *Ocyurus chrysurus* (Bloch, 1791).

- Genus *Paracaesio*

— Japanese snapper, *Paracaesio caerulea* (Katayama, 1934).

— Vanuatu snapper, *Paracaesio gonzalesi* (Fourmanoir & Rivaton, 1979).

— Saddle-back snapper, *Paracaesio kusakarii* (Abe, 1960).

— *Paracaesio paragrapsimodon* (Anderson & Kailola, 1992).

— Dirty ordure snapper, *Paracaesio sordida* (Abe & Shinohara, 1962).

— Cocoa snapper, *Paracaesio stonei* (Raj & Seeto, 1983).

— *Paracaesio waltervadi* (Anderson & Collette, 1992).

— Yellowtail blue snapper, *Paracaesio xanthura* (Bleeker, 1869).

- Genus *Parapristipomoides*

— Scalemouth jobfish, *Parapristipomoides squamimaxillaris* (Kami, 1973).

- Genus *Pinjalo*

— Slender pinjalo, *Pinjalo lewisi* (Randall, Allen & Anderson, 1987).

— Pinjalo, *Pinjalo pinjalo* (Bleeker, 1850).

- Genus *Pristipomoides*

— Wenchman, *Pristipomoides aquilonaris* (Goode & Bean, 1896).

— Ornate jobfish, *Pristipomoides argyrogrammicus* (Valenciennes, 1832).

— Goldflag jobfish, *Pristipomoides auricilla* (Jordan, Evermann & Tanaka, 1927).
— Crimson jobfish, *Pristipomoides filamentosus* (Valenciennes, 1830).
— Golden eye jobfish, *Pristipomoides flavipinnis* (Shinohara, 1963).
— Slender wenchman, *Pristipomoides freemani* (Anderson, 1966).
— Cardinal snapper, *Pristipomoides macrophthalmus* (Muller & Troschel, 1848).
— Goldbanded jobfish, *Pristipomoides multidens* (Day, 1871).
— Lavender jobfish, *Pristipomoides sieboldii* (Bleeker, 1854).
— Sharptooth jobfish, *Pristipomoides typus* (Bleeker, 1852).
— Oblique-banded snapper, *Pristipomoides zonatus* (Valenciennes, 1830).

• Genus *Randallichthys*
— Randall's snapper, *Randallichthys filamentosus* (Fourmanoir, 1970).

• Genus *Rhomboplites*
— Vermilion snapper, *Rhomboplites aurorubens* (Cuvier, 1829).

• Genus *Symphorichthys*
— Sailfin snapper, *Symphorichthys spilurus* (Gunther, 1874).

• Genus *Symphorus*
— Chinamanfish, *Symphorus nematophorus* (Bleeker, 1860).

Snipe eel

Snipe eels are a family, Nemichthyidae, of eels. They are pelagic fishes, found in the middle depths of most oceans. The common name comes from the resemblance of the jaw to the bill of a snipe.

Genera and Species

There are nine species in three genera, including:

- Genus *Avocettina*
  - — *Avocettina acuticeps* (Regan, 1916).
  - — *Avocettina bowersii* (Garman, 1899).
  - — Avocet snipe-eel, *Avocettina infans* (Gunther, 1878).
  - — *Avocettina paucipora* (Nielsen & Smith, 1978).
- Genus *Labichthys*
  - — *Labichthys carinatus* (Gill & Ryder, 1883).
  - — *Labichthys yanoi* (Mead & Rubinoff, 1966).
- Genus *Nemichthys*
  - — Slender snipe eel, *Nemichthys scolopaceus* (Richardson, 1848).
  - — Boxer snipe eel, *Nemichthys curvirostris* (Stromman, 1896).
  - — *Nemichthys larseni* (Nielsen & Smith, 1978).

Southern Flounder

The southern (or armless) flounders are a small family of flounders found in Antarctic and sub-Antarctic waters.

Their bodies are greatly compressed, with both eyes on the left side of the head. The caudal fin is separate, and the pectoral fins are rudimentary or entirely absent; none of the fins have spines. The lateral line is straight and well-developed. The family varies considerably in size, from the 6.3 cm of *Achiropsetta slavae* to the 60 cm length of the armless flounder *Neoachiropsetta milfordi*.

Most of the species have only recently been discovered, and little is known of their habits.

Species

- Genus *Achiropsetta*
  - — *Achiropsetta slavae*
  - — *Achiropsetta tricholepis*, (Finless flounder)

- Genus *Mancopsetta*
  - — *Mancopsetta maculata antarctica*, (Antarctic armless flounder)
  - — *Mancopsetta maculata maculata*
- Genus *Neoachiropsetta*
  - — Armless flounder, *Neoachiropsetta milfordi* (Penrith, 1965)
- Genus *Pseudomancopsetta*
  - — *Pseudomancopsetta andriashevi*

Spiny Dogfish

The spiny dogfish or piked dogfish, *Squalus acanthias*, is one of the best known of the dogfish, members of the family Squalidae in the order Squaliformes. There are actually several species to which the names are applied, but all are readily distinguished by their having two spines (one anterior to each dorsal fin) and their lack of an anal fin. It is found in shallow waters and offshore in most parts of the world, especially in temperate waters.

Morphology and Behaviour

The spiny dogfish has dorsal spines, no anal fin, and white spots along its back. The caudal fin has asymmetrical lobes, forming a heterocercal tail. Males mature at around 11 years of age, growing to 80-100 cm in length; females mature in 18-21 years and are slightly larger than males, reaching 100-124 cm. Both sexes are greyish brown in colour and are countershaded. Males are identified by a pair of pelvic fins modified as sperm-transfer organs, or "claspers". The male inserts one clasper into the female cloaca during copulation.

The species name *acanthias* refers to the shark's two spines. These are used defensively; if captured, the shark can arch its back to pierce its captor. Glands at the base of the spines secrete a mild poison.

The spiny dogfish forms large schools of hundreds to thousands of sharks. It is from this behaviour that their name is derived, as fisherman likened these large schools to packs of dogs. Schools are often composed entirely of sharks of like size and sex. The shark feeds on bony fishes, smaller sharks, and various invertebrates. It is also a common prey item for large fish, other sharks, and marine mammals.

Reproduction is aplacental viviparous, which was before called ovoviviparity. Fertilization is internal. The male inserts one claspers into the female oviduct orifice and injects sperm along a groove on the clasper's dorsal section. Immediately following fertilization, the eggs are surrounded by thin shells called candles, with one candle usually surrounding several eggs. Mating takes place in the winter months, with gestation lasting 22-24 months (the longest of any vertebrate). Litters range between 2 and 11 but average 6 or 7.

Commercial Use: Spiny dogfish are fished for food in Europe, the United States, Canada, New Zealand and Chile. The meat is primarily consumed in England, France, the Benelux countries and Germany. The fins and tails are processed into fin needles and are used in less expensive versions of shark fin soup in Chinese cuisine. In England it is sold in "fish and chip shops" as "rock salmon", in France it is sold as "small salmon" (saumonette) and in Belgium it is sold as "sea eel" (zeepaling). It is also used as fertilizer, liver oil, and pet food, and, because of its availability and manageable size, as a popular vertebrate dissection specimen, especially in high schools.

Sprattus

The squaretails are a genus, *Tetragonurus,* of perciform fishes, the only genus in the family Tetragonuridae. They are found in tropical and subtropical oceans, and feed on jellyfish and ctenophores.

Species

There are 3 species:

- Bigeye squaretail, *Tetragonurus atlanticus* (Lowe, 1839).
- Smalleye squaretail, *Tetragonurus cuvieri* (Risso, 1810).
- Pacific squaretail, *Tetragonurus pacificus* (Abe, 1953).

Pacific Staghorn Sculpin

The Pacific staghorn sculpin, *Leptocottus armatus,* is a common sculpin (Cottidae) found in shallow coastal waters along the Pacific coast from Alaska to Baja California. The sole member of its genus, it is unusual for having spined antler-like projections on its gill covers; it can raise the projections as a defence mechanism.

Staghorn sculpins are slender fish, with a greyish olive above, pale creamy yellow sides, and a white belly. The first dorsal fin has 7 spines and usually a dark spot in the posterior half, while the second dorsal has 17 rays. The anal fin also has 17 rays, while the pelvic fins have four rays. The fins have barred patterns of varying prominence. They can reach a length of 46 cm.

They are common in estuaries and coastal lagoons, where they feed on a variety of invertebrates, primarily amphipods such as *Corophium.*

Stargazer

The stargazers are a family Uranoscopidae of perciform fish that have eyes on top of their heads (thus the name). The family includes about 50 species in 8 genera, all marine and found worldwide in shallow waters.

In addition to the top-mounted eyes, stargazers also have a large upward-facing mouth in a large head. Their usual habit is to bury themselves in sand, and leap upwards to ambush prey (benthic fish and invertebrates) that pass overhead. Some species have a worm-shaped lure growing out of the floor of the mouth, which they can wiggle to attract prey's attention. Both the dorsal and anal fins are relatively long; some lack dorsal spines. Lengths range from 18 cm up to 90 cm, for the giant stargazer *Kathetostoma giganteum.*

Stargazers are venomous; they have two large poison spines situated behind the opercle and above the pectoral fins. They can also cause electric shocks.

Genera and Species

- Genus *Astroscopus*
- Genus *Genyagnus*
  - — Spotted stargazer, *Genyagnus monopterygius* (Schneider, 1801)
- Genus *Gnathagnus*
  - — Armoured blenny, *Gnathagnus armatus* (Kaup, 1858)
  - — *Gnathagnus cribratus* (Kishimoto, 1989)
  - — Freckled stargazer, *Gnathagnus egregius* (Jordan & Thompson, 1905)
  - — Bulldog stargazer, *Gnathagnus innotabilis* (Waite, 1904)
- Genus *Ichthyscopus*
- Genus *Kathetostoma*
  - — Lancer stargazer, *Kathetostoma albigutta* (Bean, 1892)
  - — Smooth stargazer, *Kathetostoma averruncus* (Jordan & Bollman, 1890)
  - — *Kathetostoma canaster* (Gomon & Last, 1987)
  - — Spiny stargazer, *Kathetostoma cubana* (Barbour, 1941)
  - — *Kathetostoma fluviatilis* (Hutton, 1872)
  - — Giant stargazer, *Kathetostoma giganteum* (Haast, 1873)
  - — Common stargazer, *Kathetostoma laeve* (Bloch & Schneider, 1801)
  - — Deepwater stargazer, *Kathetostoma nigrofasciatum* (Waite & McCulloch, 1915)
- Genus *Pleuroscopus*
- Genus *Selenoscopus*
- Genus *Uranoscopus*
- Genus Xenocephalus

Starry Flounder

The starry flounder (*Platichthys stellatus*) is a common flatfish found around the margins of the North Pacific.

The distinctive features of the starry flounder include the combination of black and white-to-orange bar on the dorsal and anal fins, as well as the skin covered with scales modified into tiny star-shaped plates or tubercles (thus both the common name and species epithet), resulting in a rough feel. The eyed side is black to dark brown, while the lower side is white or cream-coloured. Although classed as "righteye flounders," individuals may have their eyes on either the right or left side. They have been recorded at up to 91 cm and 9 kg.

Starry flounders are inshore fish, ranging up estuaries well into the freshwater zone, to the first riffles, with young found as much as 120 km inland. In marine environments, they occur as deep as 375 m. They glide over the bottom by rippling their dorsal and anal fins, seeking out a variety benthic invertebrates. Larvae start out consuming planktonic algae and crustaceans, then as they metamorphose they shift to larger animals.

On the western side of the Pacific they occur as far south as Japan and Korea, ranging through the Aleutian Islands, the coast of Alaska, Canada, and down the West Coast of the US as far as the mouth of the Santa Ynez River in Santa Barbara County, California. They are an important game and food fish across their range.

Stickleback

The Gasterosteidae are a family of fish including the sticklebacks. FishBase currently recognises 16 species in the family, grouped in 5 genera. However several of the species have a number of recognised subspecies, and the taxonomy of the family is thought to be in need of revision. Although some authorities give the common name of the family as "sticklebacks and tube-snouts", the tube-snouts are currently classified in the related family Aulorhynchidae.

The family includes the three-spined stickleback *Gasterosteus aculeatus aculeatus*, common in Northern Temperate Climates around the world including Europe, Alaska, and Japan and colloquially known in Britain as the "tiddler". Niko Tinbergen's studies of the behaviour of this fish were important in the early development of ethology as an example of a fixed action pattern.

Species

- Genus *Apeltes*
  - — Fourspine stickleback, *Apeltes quadracus* (Mitchill, 1815).
- Genus *Culaea*
  - — Brook stickleback, *Culaea inconstans* (Kirtland, 1840).
- Genus *Gasterosteus*
  - — Three-spined stickleback, *Gasterosteus aculeatus aculeatus* (Linnaeus, 1758).
  - — Santa Ana stickleback, *Gasterosteus aculeatus santaeannae* (Regan, 1909).
  - — Unarmoured threespine stickleback, *Gasterosteus aculeatus williamsoni* (Girard, 1854).
  - — *Gasterosteus crenobiontus* (Bacescu & Mayer, 1956).
  - — *Gasterosteus microcephalus* (Girard, 1854).
  - — Blackspotted stickleback, *Gasterosteus wheatlandi* (Putnam, 1867).
- Genus *Pungitius*
  - — *Pungitius hellenicus* (Stephanidis, 1971).
  - — *Pungitius kaibarae* (Tanaka, 1915).
  - — Smoothtail ninespine stickleback, *Pungitius laevis* (Cuvier, 1829).
  - — Southern ninespine stickleback, *Pungitius platygaster* (Kessler, 1859).
  - — Ninespine stickleback, *Pungitius pungitius* (Linnaeus, 1758).
  - — Amur stickleback, *Pungitius sinensis* (Guichenot, 1869).
  - — Sakhalin stickleback, *Pungitius tymensis* (Nikolskii, 1889).
- Genus *Spinachia*
  - — Sea stickleback, *Spinachia spinachia* (Linnaeus, 1758).

One unusual features of sticklebacks is that they have no scales, although some species have bony armour plates. They are closely related to pipefish and seahorses.

Weever

Weevers (or weever-fish) are eight species of fish of family Trachinidae, order Perciformes. They are long (up to 37 cm), mainly brown and have poisonous spines on their first dorsal fin and gills. During the day, weevers bury themselves in sand, just showing their eyes, and snatch prey as it comes past, which consists of shrimps and small fish. Weevers are unusual in not having a swim bladder as do most bony fishes and as a result sink as soon as they stop actively swimming.

This fish is used in the recipe of the bouillabaisse.

Weevers are sometimes erroneously called 'weaver fish', although the word is unrelated. In fact, the word 'weever' is believed to derive from the Old French word 'wivre', meaning serpent or dragon, from the Latin 'vipera'. It is sometimes also known as the viperfish, although it is not related to the viperfish proper. In Australia, sand perches of the family Mugiloididae are known as weevers.

Species

There are nine species in two genera:

- Genus *Echiichthys*
  - — Lesser weever, *Echiichthys vipera* (Cuvier, 1829).
- Genus *Trachinus*
  - — Spotted weever, *Trachinus araneus* (Cuvier, 1829).
  - — Guinean weever, *Trachinus armatus* (Bleeker, 1861).
  - — Sailfin weever, *Trachinus collignoni* (Roux, 1957).
  - — *Trachinus cornutus* (Guichenot, 1848).
  - — Greater weever, *Trachinus draco* (Linnaeus, 1758).
  - — Striped weever, *Trachinus lineolatus* (Fischer, 1885).
  - — Cape Verde weever, *Trachinus pellegrini* (Cadenat, 1937).
  - — Starry weever, *Trachinus radiatus* (Cuvier, 1829).

Interaction with Humans

Causes, Frequency and Prevention: Most human stings are inflicted by the lesser weever which habitually remains buried in sandy areas of shallow water and is thus more likely to come into contact with bathers than other species (such as the greater weever, which prefers deeper water), stings from other species are generally limited to anglers and commercial fishermen. Even very shallow water (sometimes little more than damp sand) may harbour lesser weevers. The vast majority of injuries occur to the foot and are the result of stepping on buried fish, other common sites of injury are the hands and buttocks.

Stings are most common in the hours before and after low tide (especially at springs) so one possible precaution is to avoid bathing or paddling at these times. They also increase in frequency during the summer (to a maximum in August) but this is probably the result of the greater number of bathers.

The lesser weever can be found from the southern North Sea to the Mediterranean and is common around the south coast of the United Kingdom, the Atlantic coast of France and Spain, and the northern coast of the Mediterranean. The high number of bathers found on popular tourist beaches in these areas means that stings are common although individual chances of being stung are low. The South Wales Evening Post stated (on 8 August 2000) that around 40 weever stings are recorded in the Swansea and Gower area every year however many victims will not seek medical assistance and go uncounted.

Weever stings have been known to penetrate wet suit boots even through a rubber sole (if thin) and it is recommended that bathers and surfers wear sandals, "jelly shoes" or wetsuit boots with a relatively hard sole and avoid sitting or "rolling" in the shallows.

Symptoms: The pain from weever stings has been described as so severe that sailors stung by the fish would cut off their stung fingers or hands in a desperate attempt to relieve the pain, this is however highly unlikely. Lifeguards on the south coasts of England and Wales deal with weever stings almost daily, and stings are often described as "extremely painful" and "much worse than a wasp [or bee] sting"—some victims find the comparison to wasp or bee stings offensive.

The following symptoms may occur following a weever sting:

- *Common/minor symptoms:* Severe pain, itching, swelling, heat, redness, numbness, tingling, nausea, vomiting, joint aches, headaches, abdominal cramps, lightheadedness, and tremors.
- *Rare/severe symptoms:* Abnormal heart rhythms, weakness, paralysis, shortness of breath, seizures, decreased blood pressure, gangrene and tissue degeneration, unconsciousness, death.

Treatment

Although extremely unpleasant, weever stings are not generally dangerous and the pain will ease considerably within a few hours even if untreated. Complete recovery may take a week or more; in a few cases victims have reported swelling and/or stiffness persisting for months after envenomation.

First Aid treatment consists of immersing the affected area in hot water (as hot as the victim can bear) which will accelerate denaturation of the protein based venom. Usual experience is that the pain then fades within ten to twenty minutes, as the water cools. Folklore often suggests the addition of substances to the hot water including urine, vinegar and epsom salts, but this is of limited (if any) value. Heat should be applied for at least 15 minutes but, as a rule of thumb: the longer the delay (before heat is applied) the longer the treatment should be continued. Once the pain has eased the injury should be checked for the remains of broken spines and any found need to be removed. Over the counter analgesics such as aspirin or ibuprofen may be of assistance in management of pain and can also reduce oedema.

Medical advice should be sought if any of the symptoms

listed above as Rare/Severe are observed, if swelling spreads beyond the immediate area of injury (e.g. from hand to arm), if symptoms persist or if any other factor causes concern. Medical treatment consists of symptom management, analgesia (often with opiates) and the same heat treatment as for first aid - more systemic treatment using anti-histamines has been largely discredited.

Fatalities

Some severe cases of poisoning may be fatal. The only recorded death in the UK occurred in 1927, when a fisherman off Dungeness suffered multiple stings. There is some suspicion that the victim may have died of other medical causes exacerbated by the stings .

Jonathan Wickings died after being stung by an unknown sea creature off the coast of Majorca in 1998. This was reported as a possible weever sting although he was not in contact with the seabed and some witnesses reported seeing a "snake" in the water. Other reports suggested that his death was the result of a fluke, whereby the venom was injected directly into a vein, causing immediate, body-wide dispersal.

Stingray

Dasyatidae is a family of rays, cartilaginous marine fishes, related to skates and sharks. Dasyatids are common in tropical coastal waters throughout the world, and there are fresh water species in Asia (*Himantura* sp.), Africa, and Florida (*Dasyatis sabina*). Most dasyatids are neither threatened nor endangered. The species of the genera Potamotrygon, Paratrygon, and Plesiotrygon are all endemic to the freshwaters of South America.

Dasyatids swim with a "flying" motion, propelled by motion of their large pectoral wings (commonly mistaken as "fins"). Their stinger is a razor-sharp, barbed, or serrated cartilaginous spine which grows from the ray's whip-like tail (like a fingernail), and can grow as long as 37 cm (about 14.6 inches).

On the underside of the spine are two grooves containing venom-secreting glandular tissue. The entire spine is covered with a thin layer of skin called the integumentary sheath, in which venom is concentrated. This gives them their common name of stingrays, but the name can also be used to refer to any poisonous ray.

Some adult rays may be no larger than a human palm, while other species, like the short-tail stingray, may have a body of six feet in diameter, and an overall length, including their tail, of fourteen feet.

Stingrays may also be called the whip-tailed rays though this usage is much less common.

A group or collection of stingrays is commonly referring to as a "fever" of stingrays.

Feeding Habits: Stingrays are flat so as to hide on the depths of the sea. They ruffle up the sand and hide beneath it. Since their eyes are on top of their body and their mouths on the bottom, stingrays cannot see their prey. Instead, they use the sense of smell and electro-receptors, similar to those of the shark. They feed primarily on molluscs and crustaceans and occasionally on small fish. Their mouths contain powerful, shell-crushing teeth. Rays settle on the bottom while feeding, sometimes leaving only their eyes and tail visible.

Stinging Mechanism

Dasyatids generally do not attack aggressively or even actively defend themselves. When threatened, their primary reaction is to swim away. However, when they are attacked by predators or stepped on, the barbed stinger in their tail is whipped up. This attack is normally ineffective against their main predator, sharks. Humans are usually stung in the foot region (depending on the size of the stingray); it is also possible, although less likely, to be stung by brushing against the stinger. The stinger often breaks off in the wound, which is non-fatal to the stingray, and will be regrown. Contact with the stinger causes local trauma (from the cut itself), pain and swelling from the venom, and possible later infection from bacteria on parts of the stinger left in the wound. Immediate injuries to humans include, but are not limited to: poisoning, punctures, severed arteries and possibly death. Fatal stings are extremely rare.

Treatment for stings includes application of near-scalding water, which helps ease pain by denaturing the complex venom protein, and antibiotics. Immediate injection of local anesthetic in and around the wound is very helpful, as is the use of adjunct opiates such as intramuscular pethidine.

Local anesthetic brings almost instant relief for several hours. Other possible pain remedies include papain (papaya extract, contained in unseasoned powdered meat tenderiser), which may break down the protein of the toxins, although this may be more appropriate for jellyfish and similar stings. Folklore incorrectly holds that one should urinate on the stung area; in actuality, urine and vinegar are not effective treatments. Pain normally lasts up to 48 hours, but is most severe in the first 30–60 minutes and may be accompanied by nausea, fatigue, headaches, fever and chills. All stingray injuries should be medically assessed; the wound needs to be thoroughly cleaned, and surgical exploration is often required to remove any barb fragments remaining in the wound. Following cleaning, a radiograph is helpful to confirm removal of all the fragments. However, not all remnants are radio-opaque; ultrasound imaging is useful in difficult cases.

Reproduction: Mating season occurs in the winter. When a male is courting a female, he will follow her closely, biting at her pectoral disc. During mating, the male will go on top of the female (his belly on her back) and put one of his claspers into her vent.

Most rays are viviparous, bearing live young in "litters" of five to ten. The female holds the embryos in the womb without a placenta. Instead, the embryos absorb nutrients from a yolk sac, and after the sac is depleted the mother provides uterine milk.

As Food: Rays may be caught on a fishing line using small crabs as bait, and are often caught accidentally; they may also be speared from above. They are edible. Small rays may be cooked similarly to other fish, typically grilled or battered and fried. While not independently valuable as a food source, the stingray's capacity to damage shell fishing grounds can lead to bounties being placed on their removal. Also, they sometimes eat squid and crabs.

Stingray recipes abound throughout the world, with dried forms of the wings being most common. For example, in Malaysia it is grilled using charcoal and is a popular dish known as Ikan Bakar. Generally, the most prized parts of the stingray are the wings, the "cheek" (the area surrounding the eyes) and the liver. The rest of the ray is considered too rubbery to have any culinary uses.

In Iceland, eating pickled stingray ("kæst skata") on December 23 is an old tradition.

Viewing: Stingrays are usually very docile creatures. The customary reaction of the stingray is to immediately flee the vicinity of a disturbance. Nevertheless, certain larger species are located in waters where they are easily excitable due to possible attacks from feeding sharks and should be approached with caution, as the stingray's defensive reflex and effort to flee may result in human contact with the stinger, resulting in serious injury or even (though rarely) death.

Dasyatids are not normally visible to swimmers, but divers and snorkellers may find them in shallow sandy waters, more so when the water is unseasonably warm.

In the Cayman Islands, there are several dive sites called Stingray City, Grand Cayman, where divers and snorkellers can swim with large southern stingrays (*Dasyatis Americana*) and feed them by hand.

There is also a "Stingray City" in the sea surrounding the Caribbean island of Antigua. It consists of a large, shallow reserve where the rays live, and snorkelling is possible.

In Belize off the island of Ambergris Caye there is a popular marine sanctuary called Hol Chan. Here divers and snorkellers often gather to watch stingrays and nurse sharks that are drawn to the area by tour operators who feed the animals.

Many Tahitian island resorts regularly offer guests the chance to "feed the stingrays and sharks". This consists of taking a boat to the outer lagoon reefs then standing in waist-high water while habituated stingrays swarm around, pressing right up against you seeking food from your hand or tossed into the water. The boat owners also "call in" sharks, which when they arrive from the ocean swoop through the shallow water above the reef and snatch food offered to them.

Most major aquariums feature stingrays, including the National Baltimore Aquarium and the Tennessee Aquarium in Chattanooga.

Where there are stingray touch tanks where visitors can "pet" rays or when show divers routinely hand feed rays in giant saltwater exhibits, for diver and visitor safety the spines on the rays are snipped off with a pair of pliers. The tip of the spine is then presented as a harmless stub that can't penetrate the skin of visitors or divers who routinely handle the docile rays.

The Atlantis Paradise Island Hotel houses many eagle rays, sting rays and one manta ray. The rays are often coexhibited with other marine life, such as the Caribbean reef shark. The Georgia Aquarium allows petting of southern stingrays in their "Georgia Explorer" exhibit. Similarly, visitors may use two fingers at a time to touch rays (with sting removed) and related guitarfish in outdoor exhibits at the Aquarium of the Pacific in Long Beach, California. Petting stingrays is also permitted in a special tank at the Blue Planet Aquarium, Ellesmere Port, England. Likewise, the Mote Marine Aquarium in Sarasota, Florida and the North Carolina Aquarium in Manteo, NC, allow visitors to pet a variety of rays in a controlled tank setting. Coral World Marine Park in St. Thomas, USVI, even allows supervised feeding of southern stingrays by visitors, as does Underwater Adventures Aquarium at the Mall of America in Bloomington, Minnesota.

In 2006, the Tampa Bay Devil Rays added a 35 foot (10.7 m), 10 000 gallon (38 000 L), touch tank in their stadium where fans get a chance to interact with dozens of rays.

Danger to Humans

He most famous stingray-related injury is the one that resulted in Steve Irwin's death. On September 4, 2006, he was pierced in the chest with a stingray spine while snorkeling in Australia. He was killed from severe loss of blood from the heart to the abdominal cavity, and was dead by the time his team had brought him to a hospital.

Recently, a Malaysian celebrity Kah Low was stung by a sting ray while surfing in the Sunset beach vicinity. On May 13, 2007 he received urgent medical care at Huntington Beach hospital and is reported to be shaken but otherwise doing well.

These chest-puncturing wounds are not typical, and most injuries that humans get from stingrays are because they step on one by accident. Stingrays are often semi-buried in the sand so they are easy to overlook. It is also possible to be bitten, but this is also a rare type of injury, since the mouth is on the bottom side of the ray. A good way to avoid stepping on a stingray while walking in shallow water is to slide your feet along the ground rather than taking steps. If a stingray feels something moving towards them, it will flee. If stepped on, however, its first reaction will be to sting.

Species

There are about seventy species in six genera:

- Genus *Dasyatis*
  - — *Dasyatis acutirostra* (Nishida & Nakaya, 1988).
  - — Red stingray, *Dasyatis akajei* (Muller & Henle, 1841).
  - — Southern stingray, *Dasyatis americana* (Hildebrand & Schroeder, 1928).
  - — Plain maskray, *Dasyatis annotata* (Last, 1987).

— Bennett's stingray, *Dasyatis bennetti* (Muller & Henle, 1841).

— Short-tail stingray or bull ray, *Dasyatis brevicaudata* (Hutton, 1875).

— Whiptail stingray, *Dasyatis brevis* (Garman, 1880).

— Roughtail stingray, *Dasyatis centroura* (Mitchill, 1815).

— Blue stingray, *Dasyatis chrysonota* (Smith, 1828).

— Diamond stingray, *Dasyatis dipterura* (Jordan & Gilbert, 1880).

— Estuary stingray, *Dasyatis fluviorum* (Ogilby, 1908).

— Smooth freshwater stingray, *Dasyatis garouaensis* (Stauch & Blanc, 1962).

— Sharpsnout stingray, *Dasyatis geijskesi* (Boeseman, 1948).

— Giant stumptail stingray, *Dasyatis gigantea* (Lindberg, 1930).

— Longnose stingray, *Dasyatis guttata* (Bloch & Schneider, 1801).

— *Dasyatis hastata* (DeKay, 1842).

— Izu stingray, *Dasyatis izuensis* (Nishida & Nakaya, 1988).

— Bluespotted stingray, *Dasyatis kuhlii* (Muller & Henle, 1841).

— Yantai stingray, *Dasyatis laevigata* (Chu, 1960).

— Mekong stingray, *Dasyatis laosensis* (Roberts & Karnasuta, 1987).

— Brown stingray, *Dasyatis latus* (Garman, 1880).

— Painted maskray, *Dasyatis leylandi* (Last, 1987).

— Longtail stingray, *Dasyatis longa* (Garman, 1880).

— Daisy stingray, *Dasyatis margarita* (Gunther, 1870).

— Pearl stingray, *Dasyatis margaritella* (Compagno & Roberts, 1984).

— *Dasyatis marianae* (Gomes, Rosa & Gadig, 2000).

— Marbled stingray, *Dasyatis marmorata* (Steindachner, 1892).

— Pitted stingray, *Dasyatis matsubarai* (Miyosi, 1939).

— Smalleye stingray, *Dasyatis microps* (Annandale, 1908).

— Multispine giant stingray, *Dasyatis multispinosa* (Tokarev, 1959).

— Blackish stingray, *Dasyatis navarrae* (Steindachner, 1892).

— Common stingray, *Dasyatis pastinaca* (Linnaeus, 1758).

— Smalltooth stingray, *Dasyatis rudis* (Gunther, 1870).

— Atlantic stingray, *Dasyatis sabina* (Lesueur, 1824).

— Bluntnose stingray, *Dasyatis say* (Lesueur, 1817).

— Chinese stingray, *Dasyatis sinensis* (Steindachner, 1892).

— Thorntail stingray, *Dasyatis thetidis* (Ogilby, 1899).

— Tortonese's stingray, *Dasyatis tortonesei* (Capape, 1975).

— Cow stingray, *Dasyatis ushiei* (Jordan & Hubbs, 1925).

— Pale-edged stingray, *Dasyatis zugei* (Muller & Henle, 1841).

• Genus *Himantura*

— Pale-spot whip ray, *Himantura alcockii* (Annandale, 1909).

— Bleeker's whipray, *Himantura bleekeri* (Blyth, 1860).

— Freshwater whipray, *Himantura chaophraya* (Monkolprasit & Roberts, 1990).

— Dragon stingray, *Himantura draco* (Compagno & Heemstra, 1984).

— Pink whipray, *Himantura fai* (Jordan & Seale, 1906).

— Ganges stingray, *Himantura fluviatilis* (Hamilton, 1822).

— Sharpnose stingray, *Himantura gerrardi* (Gray, 1851).

— Mangrove whipray, *Himantura granulata* (Macleay, 1883).

— *Himantura hortlei* (Last, Manjaji-Matsumoto & Kailola, 2006).

— Scaly whipray, *Himantura imbricata* (Bloch & Schneider, 1801).
— Pointed-nose stingray, *Himantura jenkinsii* (Annandale, 1909).
— Kittipong's stingray, *Himantura kittipongi*
— Marbled freshwater whip ray, *Himantura krempfi* (Chabanaud, 1923).
— *Himantura lobistoma* (Manjaji-Matsumoto & Last, 2006).
— Blackedge whipray, *Himantura marginatus* (Blyth, 1860).
— Smalleye whip ray, *Himantura microphthalma* (Chen, 1948).
— Marbled whipray, *Himantura oxyrhyncha* (Sauvage, 1878).
— Pacific chupare, *Himantura pacifica* (Beebe & Tee-Van, 1941).
— *Himantura pareh* (Bleeker, 1852).
— Round whip ray, *Himantura pastinacoides* (Bleeker, 1852).
— Chupare stingray, *Himantura schmardae* (Werner, 1904).
— White-edge freshwater whip ray, *Himantura signifer* (Compagno & Roberts, 1982).
— Black-spotted whipray, *Himantura toshi* (Whitley, 1939).
— Whitenose whip ray, *Himantura uarnacoides* (Bleeker, 1852).
— Honeycomb stingray, *Himantura uarnak* (Forsskal, 1775).
— Leopard whipray, *Himantura undulata* (Bleeker, 1852).
— Dwarf whipray, *Himantura walga* (Muller & Henle, 1841).

- Genus *Pastinachus*

— Cowtail stingray, *Pastinachus sephen* (Forsskal, 1775).
— *Pastinachus solocirostris* (Last, Manjaji & Yearsley, 2005).

- Genus *Pteroplatytrygon*
  - — Pelagic stingray, *Pteroplatytrygon violacea* (Bonaparte, 1832).
- Genus *Taeniura*
  - — Round stingray, *Taeniura grabata* (E. Geoffroy Saint-Hilaire, 1817).
  - — Bluespotted ribbontail ray, *Taeniura lymma* (Forsskal, 1775).
  - — Blotched fantail ray, *Taeniura meyeni* (Muller & Henle, 1841).
- Genus *Urogymnus*
  - — Porcupine ray, *Urogymnus asperrimus* (Bloch & Schneider, 1801).
  - — Thorny freshwater stingray, *Urogymnus ukpam* (Smith, 1863).

Stonefish

The stonefish, *Synanceia verrucosa,* also known as the reef stone or dornorn is a carnivorous ray-finned fish with venomous spines that lives on reef bottoms, camouflaged as a rock. It is the most venomous fish in the world..

Range: The stonefish lives primarily above the tropic of Capricorn: It is the most widespread species of the stonefishes family, and is known to be found in the shallow tropical marine waters of the Pacific and Indian oceans, ranging from the Red Sea to the Queensland Great Barrier Reef.

Description: The average length of most stonefish is about 35-50 centimetres. It has a mottled greenish to mostly brown colour which aids in its ability to camouflage itself among the rocks of many of the tropical reefs. The fish eats mostly small fish, shrimp and other crustaceans.

Habitat

Its main habitat is on coral reefs, around dull coloured plants near and about rocks, or can be found dormant in the mud or sand.

Points of Note: The primary commercial significance of the stonefish is as an aquarium pet, but they are also sold for their meat in Hong Kong markets. In addition, stonefish is also consumed in Japan as expensive sashimi cuisine. Stonefish can survive out of water for up to 20 hours.

Venom

Its dorsal area is lined with spines that release a venomous toxin. It is the most dangerous of known venomous fish and its venom causes severe pain with possible shock, paralysis, and tissue death depending on the depth of the penetration. This level can be fatal to humans if not given medical attention within a couple of hours.

Immediate first aid treatment requires the immobilisation of venom at penetration site; depending on the depth of penetration this can be achieved either by firm constrictive bandaging or by a managed tourniquet sited between wound and proximal flexure.

The venom consists of a mixture of proteins, including the hemolytic stonustoxin, the neurotoxic trachynilysin and the cardioactive cardioleputin; an antivenin is available.

Since the venom is protein based, it can be denatured by the application of a very hot compress to the injury site. Some relief can be gained from infiltrating the wound with a local anaesthetic. This is a temporary measure to reduce localised pain and shock. Medical help must be sought at the earliest opportunity. Typically, surviving victims suffer localised nerve damage occasionally leading to atrophy of adjoining muscle tissues.

There have been unproven reports of osteo-arthritic sufferers experiencing improved mobility and reduction in joint pain following envenomation episode. The responsible agent has not been identified.

The pain is said to be so bad that the victims of its sting want to cut off the affected limb. The poisonous sting of Scorpion Fish and Lionfish are said to deliver the same level of pain.

Stream Catfish

The stream catfishes are family Akysidae of catfishes found in the fresh waters of South East Asia. It includes 47 species in five genera; many species are only recently described. They are mostly small, only a few species reaching as much as 10 cm in length, *Breitensteinia insignis* being the largest known at 22 cm. There are four pairs of barbels. Parakysinae is sometimes listed as an independent family.

Striped Bass

The striped bass *Morone saxatilis* is a member of the temperate bass family native to North America but widely introduced elsewhere. Among the other names used for this species are striper bass, striped sea-bass, rock, and rockfish.

Morphology and Lifespan

The striped bass is a typical member of the Moronidae family in shape, having a streamlined, silvery body marked with longitudinal dark stripes running from behind the gills to the base of the tail. Maximum size is 200 cm (6.6 feet) and maximum scientifically recorded weight 57 kg (125 US pounds). Striped bass are believed to live for up to 30 years.

Distribution: Striped bass are found along the Atlantic coastline of North America from the St. Lawrence River into the Gulf of Mexico to approximately Louisiana. They are anadromous fish that migrate between fresh and salt water. Spawning takes place in freshwater.

They have been introduced to a number of other waters outside their natural range, including Ecuador, Iran, Latvia, Mexico, Russia, South Africa, and Turkey primarily for use as gamefish and in aquaculture.

In many of the large reservoir impoundments across the United States, striped bass have been introduced by state game and fish commissions for the purposes of recreational fishing and predator control of Gizzard Shad.

Environmental Factors

The spawning success of striped bass has been studied in the San Francisco Bay-Delta water system, with a finding that high total dissolved solids (TDS) reduce spawning. At levels as low as 200 mg/L TDS there is an observable diminution of spawning productivity.

Life Cycle

Striped bass breed in freshwater and spend their adult lives in saltwater (i.e., it is anadromous). They can also live exclusively in freshwater and currently flourish in inland water bodies such as Lake Murray, Lake Powell, Lake Havasu, Lake Texoma and Lake Mead. For saltwater striped bass, four important bodies of water with breeding stocks of striped bass are: Chesapeake Bay, Massachusetts Bay/Cape Cod, Hudson River and Delaware River. There are many smaller breeding areas that contribute to the overall striped bass population such as the Takanasse Lake. It is believed that many of the rivers and tributaries that emptied into the Atlantic, had at one time, breeding stock of striped bass. One of the largest breeding areas is the Chesapeake Bay, where populations from Chesapeake and Delaware bays have intermingled.

Hybrids with other Bass

Striped bass have also been hybridised with white bass to produce sunshine bass, palmetto bass, or wiper; white perch to produce Virginia bass or Maryland bass; and yellow bass to produce paradise bass. These hybrids have been stocked in many freshwater areas across the US.

Fishing for Striped Bass

This fish is found all along the Atlantic coast, from Florida to Nova Scotia, and are caught as far north as Hudson Bay. An anadromous fish, it inhabits rivers, bays, inlets, estuaries, and creeks. It is quite abundant in the Chesapeake Bay and its tributaries. There, it frequently grows over four feet in length and weighs over 22 kg (50 lb). The largest striped bass ever caught by angling was a 35.6 kg (78.5 lb) specimen taken in

Atlantic City, NJ on September 21, 1972. The striped bass will swim up rivers a hundred miles or more, and in Maine they are quite plentiful in the Penobscot River and Kennebec River. Further south in Connecticut some very large ones are taken both offshore and in the Connecticut River, and the waters surrounding New York City have proven a fertile fishing ground with good sized specimens being caught during spring and summer months.

East Coast striped bass are typically found from the Carolinas to Nova Scotia. The Chesapeake Bay is the major producer area for striped bass, with the Hudson river being a secondary producer. Spawning migration begins in March when the migratory component of the stock returns to their natal rivers to spawn.

It is believed that females migrate after age five, these fish are believed to remain in the ocean during the spawning run males as young as two years old have been encountered in the spawning areas of the Cheasapeake Bay. The migratory range of the northern (hudson stock) extends from the Carolinas to New York's Hudson River in the winter time and from New Jersey through Maine in summertime with the greatest concentration between Long Island, New York, Rhode Island, and Massachusetts. The migration of the northern stock to the south often begins in September from areas in Maine.

On the West Coast, stripers are found throughout the San Francisco Bay and surrounding coastline. They are also found in the California Aqueduct canal system, and many California lakes. The record striped bass catch at Pyramid Lake in The Grapevine was 19 kg (42 lb). Frequent "boils" or swarms of these fish may be observed in these lakes, representing an excellent fishing opportunity, especially with *Pencil Poppers* or other similar trout-looking surface lures.

In winter they keep to their haunts, and do not go into deep water like other fish of similar habits. In the spring of the year the striped bass runs up the rivers and into other fresh water places to spawn—and then again late in the fall to shelter. The fall run is the best. They can be caught however nearly all the year round, and of all sizes.

Striped bass can be caught using a number of baits including: clams, eels, anchovies, bloodworms, nightcrawlers, chicken livers, menhaden, herring, shad, and sandworms. At times, striped bass can be very choosy about the baits they take. Because of the wide variety of baits that are known to work and their finicky nature, they are considered among fishermen as being an opportunistic or "lazy" feeder. However, it is estimated that 90 per cent of their diet is fish.

Surfcasting

Fishing from the shore is a popular method of striped bass fishing among anglers who may not have access to a boat or simply prefer to stay on shore. Shore fishing can include fishing the shores of inland waterways, saltwater ponds, rivers, and bays. Various methods of light tackle to heavy gear can be used. More challenging shore fishing along the immediate ocean coastline is often referred to as *surfcasting*. Surfcasters typically gear up a little differently than inland shore anglers as the conditions tend to be more severe, featuring high winds and heavy surf.

In addition to rod, reel, and tackle, the surfcaster's typical equipment list should include items for safety and for comfort such as waders secured by a tight wader belt to prevent filling with water, dry top, line clippers, pliers, hook cutters, and knife as well as a neck light or headlight for use at night. Additional items for safety may include steel-studded soles attached to the wader boots to improve traction, and an inflatable life vest to prevent drowning accidents in more severe conditions, as several surfcasting fatalities occur annually.

More extreme surfcasting may entail climbing on rocks far from shore to gain an advantaged position or in some cases; anglers may don wetsuits to swim to rocks in water unreachable by wading. Surfcasting gear usually includes spinning or conventional reels on rods in the 2.4-3.6 m (8-12 foot) range using lines of 7-9 kg (15-30 lb) test monofilament or equivalent diameters of braid.

Some surf-fisherman don't like to use braids because it will cut easily on rocks but recent advancements in braid are making it more acceptable in the fishing community. Braid has the advantages of a lower diameter to breaking strength than mono meaning better casting and no stretch meaning better bite detection.

High vis line is best in blitz situations when it is important to see your line. Plastic lures such as bombers, redfins, yozuris all work. When choosing a lure, the profile of the fish you are trying to mimic and the movements of the lure are more important than the colour - striped bass do not have the same rods and cones as a human eye.

Other lure choices are wooden lures, lead jigs, and soft plastics. Live bait is very effective such as herring and eels. Cut bait like chunks of herring and mackerel work well when live bait can't be fished. In the Atlantic, Striped Bass heavily pursue schools of Atlantic menhaden or more commonly known as Mossbunker. When cut up into chunks, this can be one of the most effective baits. Some other important bait choices include clams, worms, and crabs. Often a sand spike is used when fishing the surf to hold a rod fished with bait. Lead weight can be used to keep the bait to the ocean floor,

Trolling

Trolling for bass is excellent sport, and is practised a good deal by amateurs. The tackle employed is a strong hand line, and artificial bait is used with good success. This consists of silver plated spoons, bucktales with plastic trailers, and surgical tubes (representing eels). Squid and eel are also an excellent bait for trolling. In order to fasten a squid to a hook, the squid's "spine" should be pulled out and the line threaded through the 'hoods' cavity with a needle. Freshwater stripers can be caught using alewives and other shads, threadfins, crayfish, and trout. The striped bass will readily eat anything that moves, including smaller individuals of its own species.

It is a temperature-specific fish, with an optimal water temperature of 17°C (63°F). In searching for prime striper fishing grounds, focus on optimal water temperature rather than the structure of the environment. The bigger fish are more affected by water temperature than the smaller ones. The bigger fish are often large and lazy, and can be caught on cutbait since they sometimes wait for scraps missed by the smaller, faster fish, instead of using their energy to chase down their meals. Another good way to catch rockfish while trolling is try to use a 20-30 cm (8-12 in) white worm with a twirl tail depending on the size of rockfish you are going for.

Striped Burrfish

The burrfish or spiny Boxfish, *Chilomycterus Schoepfi*, is a member of the porcupinefish family Diodontidae.

Description: It is distinguished from the porcupinefish by the shorter, less sharply pointed, and immovable spines which cover the somewhat spherical body. It can inflate its body by taking either air or water into a ventral extension of the stomach. Its colour is olivaceous or brownish above and pale yellow below. The back and sides are irregularly striped with brownish, dusky, or black lines which are parallel to each other and which run obliquely downward. There are several large black spots on the sides, one just below the dorsal fin, and another behind the pectoral fin. Its maximum size is about 10 inches.

Distribution: It is found mostly in the tropics from the West Indies to Floridia, and is found sparingly along the Atlantic coast, sometimes as far north as Cape Cod, and regularly during the late summer and fall in the vicinity of New York.

Habitat

It spawns off New Jersey in July.

Diet: It feeds on invertebrates such as oysters, barnacles, and hermit crabs.

Fishing Technique

It has no commercial value but is occasionally taken by fishermen and stuffed as curio. It is often washed ashore along the beach and is a treacherous object if stepped on by mistake with bare feet. Other species of burr and porcupine fish are found in the tropics.

Sturgeon

Sturgeon is a term for a genus of fish (*Acipenser*) of which 26 species are known. One of the oldest genera of fish in existence, they are native to European, Asian, and North American waters. Sturgeons ranging from 7 -12 feet in length are by no means scarce, and some species grow up to 14 feet.

Sturgeon are bottom-feeders. With their projecting wedgeshaped snout they stir up the soft bottom, and by means of their sensitive barbels detect shells, crustaceans and small fish, on which they feed. Having no teeth, they are unable to seize larger prey.

Only a few of the species are exclusively confined to fresh water. Sturgeon are confined to the Northern Hemisphere and do not inhabit tropical regions.

In Russia the fisheries are of immense value. Early in summer the fish migrate into the rivers or towards the shores of freshwater lakes in large shoals for breeding purposes. The ova are very small, and so numerous that one female has been calculated to produce about three million in one season. The ova of some species have been observed to hatch within very few days after exclusion.

In Sturgeons that have attained maturity their growth appears to be much slower, although continuing for many years. Frederick the Great placed a number of them in the Garder See Lake in Pomerania about 1780; some of these were found to be still alive in 1866. Professor von Baer also states, as the result of direct observations made in Russia, that the Hausen (*Acipenser huso*) attains to an age of 100 years, but can live over 210 years.

In countries like England, where few sturgeons are caught, sturgeon is included as a royal fish in an act of King Edward II, although it probably only rarely graces the royal table of the present period, or even that of the lord mayor of London, who can claim all sturgeons caught in the Thames above London Bridge. Where sturgeons are caught in large quantities, as on the rivers of southern Russia and on the great lakes of North America, their flesh is dried, smoked or salted.

The ovaries, which are of large size, are prepared for caviar, for this purpose they are beaten with switches, and then pressed through sieves, leaving the membranous and fibrous tissues in the sieve, whilst the eggs are collected in a tub.

The quantity of salt added to them before they are finally packed varies with the season, scarcely any being used at the beginning of winter. Finally, one of the best sorts of isinglass is manufactured from the airbladder. After it has been carefully removed from the body, it is washed in hot water, and cut open in its whole length, to separate the inner membrane, which has a soft consistency, and contains 70 per cent of glutin.

Sturgeon (and, therefore also the caviar trade) are under severe threat from overfishing, poaching and water pollution.

Species

The twenty-six species of sturgeons (*Acipenser* and *Huso*) are nearly equally divided between the Old and New Worlds. Most are now considered to be critically endangered, endangered or vulnerable. The more important are the following (from 1911 Encyclopaedia Britannica):

- The Common sturgeon (*Acipenser sturio*), also known as the European or Baltic strugeon occurs on all the coasts of Europe, but is absent in the Black Sea. Almost all the British specimens of sturgeon belong to this species; it crosses the Atlantic and is not rare on the coasts of North America. It reaches 12 ft (4 m) long, but is always caught singly or in pairs, so that it cannot be regarded as a fish of commercial importance. The form of its snout varies with age (as in the other species), being much more blunt and abbreviated in old than in young examples. There are 11–13 bony shields along the back and 29–31 along the side of the body. The European or Atlantic sturgeon is now mostly gone from overfishing.

- The Russian sturgeon, *Acipenser gueldenstaedtii* is one of the most valuable species of the rivers of Russia, where it is known under the name Ossetra; it is said to inhabit the Siberian rivers also, and to range eastwards as far as Lake Baikal. It was so abundant in the rivers of the Black and Caspian seas that more than one-fourth of the caviares and isinglass manufactured in Iran and Russia was derived from this species. However, due to poaching and overfishing it is now a threatened species.
- The Starry sturgeon, *Acipenser stellatus*, the "Sevruga" (Ñaâðþaa) of the Russians, occurs likewise in great abundance in the rivers of the Black Sea and of the Sea of Azov. It has a remarkably long and pointed snout, like the sterlet (below), but simple barbels without fringes. Though growing only to about half the size of the preceding species, it is of no less value, its flesh being more highly esteemed, and its caviare and isinglass fetching a higher price. In 1850 it was reported that more than a million of this sturgeon are caught annually.
- The Lake sturgeon, *Acipenser rubicundus* (today: *Acipenser fulvescens*), with which, in the opinion of American ichthyologists, the sea-going sturgeon of the rivers of eastern North America, *Acipenser maculosus*, is identical, has of late years been made the object of a large and profitable industry at various places on Lake Michigan and Lake Erie; the flesh is smoked after being cut into strips and after a slight pickling in brine; the thin portions and offal are boiled down for oil; nearly all the caviare is shipped to Europe. One firm alone uses from ten thousand to eighteen thousand sturgeons a year, averaging 50 lb (23 kg) each. The sturgeons of the lakes

are unable to migrate to the sea, whilst those below Niagara Falls are great wanderers; and it is quite possible that a specimen of this species said to have been obtained from the Firth of Tay was really captured on the coast of Scotland.

- The Beluga sturgeon, *Huso huso*, the "Hausen" of Germany, is recognised by the absence of osseous scutes on the snout and by its flattened, tape-like barbels. It is one of the largest species, reaching in exceptional cases enormous lengths of more than 5m and a weight of more than 2000 lb (900 kg). It inhabits the Caspian and Black seas, and the Sea of Azov, whence in former years large shoals of the fish entered the large rivers of Russia and the Danube. But its numbers have been much thinned, and specimens of 1200 lb (540 kg) in weight have now become scarce. Its flesh, caviare and air-bladder are of a greater value than those of the smaller, more common, kinds.
- The sterlet, *Acipenser ruthenus*, is one of the smaller species, which likewise inhabits both the Black and Caspian seas, and ascends rivers to a greater distance from the sea than any of the other sturgeons; thus, for instance, it is not uncommon in the Danube at Vienna, but specimens have been caught as high up as Ratisbon and Ulm. It is more abundant in the rivers of Russia, where it is held in high esteem on account of its excellent flesh, contributing also to the best kinds of caviare and isinglass. As early as the 18th century attempts were made to introduce this valuable fish into the Province of Prussia and Sweden, but without success. The sterlet is distinguished from the other European species by its long and narrow snout and fringed barbels. It rarely exceeds a length of 3 ft (1 m).
- The family Acipenseridae includes three other genera, *Scaphirhynchus*, the shovel-headed or shovel-nosed sturgeons, distinguished by the long, broad and flat snout, the suppression of the spiracles, and the union of the longitudinal rows of scales posteriorly. All the species are

confined to fresh water. One of them is rare in the Mississippi and other rivers of North America, the other three occur in the larger rivers of eastern Asia; the beluga sturgeons of genus *Huso*, and the false shovel-headed sturgeons, of *Scaphirhynchus* sister genus, *Pseudoscaphirhyncus*, and are confined to Northeastern Asia.

Other species include:

- *Acipenser baerii*
  - — Siberian sturgeon, *Acipenser baerii baerii*
  - — Baikal sturgeon, *Acipenser baerii baicalensis*
  - Shortnose sturgeon, *Acipenser brevirostrum*
  - Yangtze sturgeon (or Dabry's sturgeon), *Acipenser dabryanus*
  - Green sturgeon, *Acipenser medirostris*
  - Sakhalin sturgeon, *Acipenser mikadoi*
  - Japanese sturgeon, *Acipenser multiscutatus*
  - Adriatic sturgeon, *Acipenser naccarii*
  - Fringebarbel sturgeon (or bastard sturgeon, ship sturgeon, spiny sturgeon, thorn sturgeon), *Acipenser nudiventris*
- *Acipenser oxyrinchus*
  - — Gulf sturgeon, *Acipenser oxyrinchus desotoi*
  - — Atlantic sturgeon, *Acipenser oxyrinchus oxyrinchus*
- Persian sturgeon, *Acipenser persicus*
- Amur sturgeon, *Acipenser schrenckii*
- Chinese sturgeon, *Acipenser sinensis*
- White Sturgeon, *Acipenser transmontanus*
- Kaluga, *Huso dauricus*

Sunfish

Several kinds of things are named "sunfish" (or "sun-fish"), including:

- Sunfish (dinghy), the International Sunfish Class sailing dinghy

- Various fish:
  - — Marine fishes:
    - ⇒ The Ocean Sunfish (*Mola mola*)
    - ⇒ The Oblong Sunfish (*Ranzania laevis*)
    - ⇒ Opah (family Lampridae; two species)
    - ⇒ The Monodactylidae (moonfishes) of genus *Selene* (order Perciformes)
    - ⇒ The African pompano (*Alectis ciliaris*) (order Perciformes)
    - ⇒ The American Gamer (*Humanis Ventrillos*) (order Tildes Athes)
    - ⇒ The spotted eagle ray (*Aetobatus narinari*), a ray (order Rajiformes)
  - — Freshwater fishes:
    - ⇒ In a technical sense, all members of family Centrarchidae (order Perciformes)
    - ⇒ The various species of genus *Lepomis* in that family, especially the bluegill and pumpkinseed
    - ⇒ A number of similar-looking Perciformes of other families, including the pygmy sunfishes
- The Banded sunfish (*Enneacanthus obesus*)

Saccopharyngiformes

Saccopharyngiformes is an order of unusual ray-finned fish superficially similar to eels, but with many internal differences. Most of the fish in this order are deep-sea types known from only a handful of specimens such as the Umbrella Mouth Gulper Eel. Saccopharyngiformes are also bioluminescent in several species.

Saccopharyngiforms lack several bones, such as the symplectic bone, the bones of the opercle, and ribs. They also have no scales, pelvic fins, or swim bladder. The jaws are quite large, and several types are notable for being able to consume fish larger than themselves.

Their myomeres (muscle segments) are V-shaped instead of W-shaped like in all other fish, and their lateral line has no pores, instead it is modified to groups of elevated tubules.

Classification

There are four families in the order:

- Cyematidae (bobtail snipe eels)
- Eurypharyngidae (pelican eel)
- Monognathidae
- Saccopharyngidae (swallowers, gulpers or gulper eels)

Diet

The gulper eel eats fish, copepods, shrimp, and plankton. It uses its mouth like a net by opening its large mouth and swimming at its prey.

Swamp eel

The swamp eels (also written "swamp-eels") are a family (Synbranchidae) of freshwater eel-like fishes of the worldwide tropics. The family includes about 18 species in four genera.

These fish are almost entirely finless; the pectoral and pelvic fin are absent, the dorsal and anal fins are vestigial, reduced to rayless ridges, and the caudal fin ranges from small to absent, depending on species.

Almost all of the species lack scales. The eyes are small, and in some cave-dwelling species they are beneath the skin, and the fish is blind.

The gill membranes are fused, and the gill opening is either a slit or pore underneath the throat. Swim bladder and ribs are also absent.

The marbled swamp eel *Synbranchus marmoratus* has been recorded at up to 150 cm in length, while the Bombay swampeel *Monopterus indicus* reaches no more than 8.5 cm.

Most of the species can breathe air. Many also burrow, and are found in the mud underneath a dried-up pond.

Sweeper

Sweepers are small, tropical marine (occasionally brackish) perciform fish of the family Pempheridae. Found in the western Atlantic Ocean and Indo-Pacific region, the family contains approximately 26 species in two genera. One species (*Pempheris xanthoptera*) is the target of subsistence fisheries in Japan, where the fish is much enjoyed for its taste. Sweepers are occasionally kept in the marine aquarium.

Physical Description

Deeply keeled, compressed bodies and large eyes typify sweepers, their form somewhat like hatchetfish; both cycloid and ctenoid scales may be present. The small, short dorsal fin begins before the body's midpoint and may have 4-7 spines; the anal fin is extensive and usually has 3 spines. The mouth is subterminal and strongly oblique. Species of the genus *Parapriacanthus* have much more cylindrical bodies.

Some species possess photophores. All but the curved sweeper (*Pempheris poeyi*) possess a gas bladder. The largest species is the common bullseye (*Pempheris multiradiata*) at 28 centimetres in length; most other species measure 16 centimetres or less. Colouration is relatively subdued.

Behaviour

Characteristically shallow water, schooling fish (especially as juveniles), sweepers are nocturnal and seek shelter under ledges or in the caves, nooks and crannies of reefs or eroded, rocky shorelines during the day. They are often found sharing these hiding places with cardinalfishes and bigeyes, fellow nightowls. At night, sweepers forage for zooplankton, their primary food items.

At least one species, the small-scale bullseye (*Pempheris compressa*) of Australia, is known to enter coastal estuaries whilst young.

Species

There are 26 species in two genera:

- Genus *Parapriacanthus*
  - — *Parapriacanthus dispar* (Herre, 1935).
  - — Slender bullseye, *Parapriacanthus elongatus* (McCulloch, 1911).
  - — *Parapriacanthus marei* (Fourmanoir, 1971).
  - — Pigmy sweeper, *Parapriacanthus ransonneti* (Steindachner, 1870).
- Genus *Pempheris*
  - — New Zealand bigeye, *Pempheris adspersa* (Griffin, 1927).
  - — Dusky sweeper, *Pempheris adusta* (Bleeker, 1877).
  - — Black-tipped bullseye, *Pempheris affinis* (McCulloch, 1911).
  - — Bronze sweeper, *Pempheris analis* (Waite, 1910).
  - — Small-scale bullseye, *Pempheris compressa* (White, 1790).
  - — *Pempheris japonica* (Doderlein, 1883).
  - — Klunzinger's bullseye, *Pempheris klunzingeri* (McCulloch, 1911).
  - — Black-edged sweeper, *Pempheris mangula* (Cuvier, 1829).
  - — *Pempheris molucca* (Cuvier, 1829).
  - — Common bullseye, *Pempheris multiradiata* (Klunzinger, 1880).
  - — *Pempheris nyctereutes* (Jordan & Evermann, 1902).
  - — *Pempheris ornata* (Mooi & Jubb, 1996).
  - — *Pempheris otaitensis* (Cuvier, 1831).
  - — Silver sweeper, *Pempheris oualensis* (Cuvier, 1831).
  - — Curved sweeper, *Pempheris poeyi* (Bean, 1885).
  - — *Pempheris rapa* (Mooi, 1998).
  - — Glassy sweeper, *Pempheris schomburgkii* (Muller & Troschel, 1848).
  - — *Pempheris schreineri* (Miranda-Ribeiro, 1915).
  - — Black-stripe sweeper, *Pempheris schwenkii* (Bleeker, 1855).

— Vanikoro sweeper, *Pempheris vanicolensis* (Cuvier, 1831).
— *Pempheris xanthoptera* (Tominaga, 1963).
— *Pempheris ypsilychnus* (Mooi & Jubb, 1996).

Swordfish

Swordfish (*Xiphias gladius*) are large, highly migratory, predatory fish characterised by a long, flat bill. They are a popular sport fish, though elusive. Swordfish are elongated, round-bodied, and lose all teeth and scales by adulthood. They reach a maximum size of 14.75 ft (4.3 m) and 3,190 lb (1,446kg). The International Game Fish Association's all-tackle angling record for a swordfish was a 3,183 lb (1,443 kg) specimen taken off Chile in 1953. They are the sole member of their family Xiphiidae.

Swardfish

Physiology

The swordfish is also known as The Gladiator (gladius) because of the sharp, sword-like bill it possesses as an addition to its streamlined physique, allowing it to cut through the water with great ease and agility. Contrary to belief the "sword" is not used to spear, but instead may be used to slash at its prey in order to injure the prey animal, to make for an easier catch. Mainly the swordfish relies on its great speed and agility in the water to catch its prey. One possible defensive use for the sword-like bill is for protection from its few natural predators. The shortfin mako shark is one of the rare sea creatures big enough and fast enough to chase down and kill an adult swordfish, but they don't always win. Sometimes in the struggle with a shark a swordfish can kill it by ramming it in the gills or belly.

Females grow larger than males, with males over 300 lb (135 kg) being rare. Females mature at 4-5 years of age in northwest Pacific while males mature first at about 3 to 4 years. In the North Pacific, batch spawning occurs in water warmer than 24 °C from March to July and year round in the equatorial Pacific. Adult swordfish forage includes pelagic fish including small tuna, dorado, barracuda, flying fish, mackerel, as well as benthic species of hake and rockfish. Squid are important when available. Swordfish likely have few predators as adults although juveniles are vulnerable to predation by large pelagic fish.

While swordfish are cold-blooded animals, they have special organs next to their eyes to heat their eyes and also their brain. Temperatures of 10 to 15 C° above the surrounding water temperature have been measured. The heating of the eyes greatly improves the vision, and subsequently improves their ability to catch prey. Out of the 25 000 species of bony fish, only about 22 are known to have the ability to heat selected body parts above the temperature of the surrounding water. These include the swordfish, marlin, tuna and some sharks.

Reproduction

Swordfish have been observed spawning in the Atlantic Ocean, in water less than 250 ft. (75 m) deep. Estimates vary considerably, but females may carry from 1 million to 29 million eggs in their gonads. Solitary males and females appear to pair up during the spawning season. Spawning occurs year-round in the Caribbean Sea, Gulf of Mexico, the Florida coast and other warm equatorial waters, while it occurs in the spring and summer in cooler regions. The most recognised spawning site is in the Mediterranean, off the coast of Italy. The height of this well-known spawning season is in July and August, when males are often observed chasing females. The pelagic eggs are buoyant, measuring 1.6-1.8mm in diameter. Embryonic development occurs during the 2 ½ days following fertilization. As the only member of its family, the swordfish has unique-looking larvae. The pelagic larvae are 4 mm long at hatching and live near the surface. At this stage, body is only lightly pigmented. The snout is relatively short and the body has many distinct, prickly scales. With growth, the body narrows. By the time the larvae reach half an inch long (12 mm), the bill is notably elongated, but both the upper and lower portions are equal in length. The dorsal fin runs the length of the body. As growth continues, the upper portion of the bill grows proportionately faster than the lower bill, eventually producing the characteristic prolonged upper bill. Specimens up to approximately 9 inches (23 cm) in length have a dorsal fin that extends the entire length of the body. With further growth, the fin develops a single large lobe, followed by a short portion that still reaches to the caudal peduncle. By approximately 20 inches (52 cm), the second dorsal fin has developed, and at approximately 60 inches (150 cm), only the large lobe remains of the first dorsal fin.

Swordfish is a particularly popular fish for cooking. Since swordfish are large animals, meat is usually sold as steaks, which are often grilled. The colour of the flesh varies by diet, with fish caught on the east coast of North America often being rosier.

However, many sources including the United States Food and Drug Administration warn about potential toxicity from high levels of methylmercury in swordfish. The FDA recommends that women who are pregnant or who may become pregnant should eat no more than one seven-ounce serving a month; others should eat no more than one serving a week.

Harvest

Swordfish were harvested by a variety of methods at small scale until the global expansion of longline fisheries in the 1950-60s. Today in the United States, swordfish are caught by harpoons, handlines, and trawling but the vast majority of catches are on pelagic longlines. In 2003, sport fishermen in the US caught 48.3mt of swordfish, approximately 4.3 per cent of the longline catch.

Longline gear can be targeted to a variety of fish, but bycatch remains a significant problem.

Are Swordfish Endangered?

Swordfish are not an endangered species. In 1998, the Natural Resourses Defence Council and SeaWeb hired Fenton Communications to conduct an advertising campaign to promote their assertion that the Swordfish population was in danger due to its popularity as a restaurant entree.

The resulting "Give Swordfish a Break" promotion was wildly successful, with 750 prominent US chefs agreeing to remove North Atlantic swordfish from their menus, and also persuaded many supermarkets and consumers across the country. The advertising campaign was repeated by the national media in hundreds of print and broadcast stories, as well as extensive regional coverage. It earned the Silver Anvil award from the Public Relations Society of America as well as Time magazine's award for the top five environmental stories of 1998.

Subsequently, the National Marine Fisheries Service proposed a swordfish protection plan that incorporated the campaign's policy suggestions. Then-President Clinton called for a ban on the sale and import of swordfish and in a landmark decision by the federal government, 132,670 square miles of the Atlantic ocean were placed off-limits to fishing as recommended by the sponsors.

Currently: In the North Atlantic, the Swordfish stock is nearly rebuilt, but biomass remains slightly below that at which MSY is produced, and abundance is increasing. This stock is considered a moderate conservation concern until the stock is fully rebuilt. There are no robust stock assessments for swordfish in the northwestern Pacific or South Atlantic, and there is a paucity of data concerning stock status in these regions. These stocks are considered unknown and a moderate conservation concern. The southwestern Pacific stock is a moderate concern due to model uncertainty, increasing catches, and declining CPUEs (catch per unit effort). Overfishing is likely occurring in the Indian Ocean, and fishing mortality exceeds the maximum recommended level in the Mediterranean, thus these stocks are considered of high conservation concern.

Green Swordtail

The Green swordtail (*Xiphophorus hellerii*) is a species of freshwater fish in family Poecilidae of order Cyprinodontiformes. It is also called Red swordtail. A live-bearer, it is closely related to the southern platyfish or "platy" (*X. maculatus*) and can interbreed with it. It is native to an area of North and Central America stretching from Veracruz, Mexico, to northwestern Honduras.

The male green swordtail grows to a maximum overall length of 14cm (5.5in) and the female to 16 cm (6.3 in). The name "swordtail" derives form the elongated lower lobe of the male's caudal fin (tailfin). Sexual dimorphism is moderate, with the female being larger than the male but lacking the "sword". The wild form is olive green in colour, with a red or brown lateral stripe and speckles on the dorsal and, sometimes, caudal fins. The male's "sword" is yellow, edged in black below. Captive breeding has produced many colour varieties, including black, red, and many patterns thereof, for the aquarium hobby.

The green swordtail prefers swift-flowing, heavily-vegetated rivers and streams, but is also found in warm springs and canals. Omnivorous, its diet includes both plants and small crustaceans, insects, and annelid worms.

X. hellerii has become a nuisance pest as an introduced species in a number of countries. It has caused ecological damage because of its ability to rapidly reproduce in high numbers. Feral populations have established themselves in southern Africa, including Natal and eastern Transvaal in South Africa and Lake Otjikoto in Namibia.

One of the most popular tropical aquarium fish, the green swordtail has been bred into various hybrid forms for the aquarium hobby due to its hardiness and suitability for community tanks.

The green swordtail, as the most common of the swordtail species (and in recognition of the fact that many captive-bred colour varieties are not green), is typically known simply as the swordtail in the aquarium hobby. It is often designated *X. helleri* (with one *i*), but authorities consider this an orthographic error and the spelling with two is the valid specific epithet. Due to interbreeding with the southern platyfish or "platy" most "swordtail" in the aquarium are hybrids to some degree.

The males' elongated caudal fins have been found to significantly affect their chances at mating. The presence of a well-endowed male spurs the maturity of females while it inhibits the maturity of juvenile males in the vicinity as the well-endowed male.

Tadpole Cod

The Tadpole cod (*Guttigadus globosus*) is a deepwater fish found in the oceanic islands off New Zealand and in the mid South Atlantic at depths ranging from 1200 - 1600 m.

The tadpole cod is a member of the family Moridae, the Morid cods, related to the true cods (of genus *Gadus*, family *Gadidae*). Like the familiar Atlantic cod, it has small whiskers on its mouth (barbels), which is presumably why it was given the name cod. Specimens have been measured up to 18 cm. It has no commercial value, and is not currently believed to be endangered. Not much is known about this species, being discovered in 1986.

The tadpole cod is distinct from the tadpole fish, *Raniceps raninus*. This fish was formerly classified in the family Ranicipitidae, called the Tadpole cods, as its sole member. The family was placed in a different order of fish, the Ophidiiformes (cusk-eels and brotulas). However it is now regarded as a member of the same family as cods, Gadidae.

Tang

Surgeon Fish are a group of saltwater fish from 6 related genera (*Acanthurus, Ctenochaetus, Naso, Paracanthurus, Zebrasoma* and *Prionurus*) which inhabit shallow reefs and beds of seagrass from the east coast of Africa northward to the Red Sea, and over the broad span of the Atlantic and Indo-Pacific oceans. The ideal temperature of their habitat is between the middle 70s and low 80s Fahrenheit (25 to 30°C).

The Zebrasoma comprise seven species of pointed-snout, disc-like bodied, sail-like finned, single peduncular-spined fishes. The terms Tang & Surgeon are often—erroneously—used interchangeably when referring to certain members of the 6 genera, however only the Zebrasoma genus can be correctly termed "Tang". For example, *Ctenochaetus Tominiensis* is often attributed the common names "Tomini Tang" or "Goldrush Tang", when in actuality it belongs to an entirely different genus. As far as common names mean anything at all, a correct one would be "Tomini Surgeon".

Tangs are usually placid community fish, many forming large groups or shoals, although fiercely territorial against other tangs or similar shaped or coloured fish in an aquarium setting. All tangs and surgeon fish are characterised by having a forward facing spine (two on some species of surgeon) on each side of the caudal peduncle. This spine, apart from terminating in a point, has angular edges that result in a scalpel-like sharpness along its length also. When alarmed, tangs can erect the spines sideways and slash at their opponent with rapid sideways movements of the tail. In some surgeons, the spines are fixed in the offensive position. The family name reflects this nature, meaning, literally, "Thorn Tail".

Care must be exercised by aquarists should it be necessary to handle these fish. An encounter with even a small, frightened tang can result in a wound requiring stitches. Large specimens are quite capable of inflicting deep, life-threatening injuries - an adult Sohal Surgeon may have spines 4cm long, quite capable of reaching the femeral arteries of an unfortunate diver.

There is some evidence to support the claim that in this fish in particular, the spines are either venom tipped or coated with epithelial cells (as in some urchins) which further complicate any injury. Colloquially described as a "Fish with a flick-knife", due mainly to their calm nature, such attacks on humans are rare. Aggression in tangs is usually confined to presenting their tails to a would be predator or attacker and is usually minimal against conspecifics, some however have noticed the purple tang, clown tang, spotted unicorn tang, and sohal tang to be more aggressive than most. Most aggression however is due to territory disputes.

Tangs are also very sensitive to disease in the home aquarium. It is usually necessary to quarantine the animals using copper sulphate or formalin for a period of around 2 weeks.

Adults range from 6 to 15 inches in length and most grow quickly even in aquariums. When considering a tang for an aquarium it is important to consider the size to which these fish can grow. Larger species such as the popular regal tang (of *Finding Nemo* fame), naso or lipstick tang, clown and sohal tangs can grow to 15 inches and require swimming room and hiding places.

Many also suggest adding aggressive tangs to the aquarium last as they are territorial and may fight and possibly kill other fish.

Tangs primarily graze on macroalgae, such as caulerpa and gracilias, although they have been observed in an aquarium setting to eat meat-based fish foods. A popular technique for aquarists, is to grow macroalgae in a sump or refugium. This technique not only is economically beneficial, but serves to promote enhanced water quality through nitrate absorption. The growth of the algae can then be controlled by feeding it to the tang.

Tarpon

The tarpons (Spanish: *sabalos*) are large coastal fish prized by anglers. They grow up to 8 feet in length and sometimes weigh 200 pounds. When swimming in oxygen poor water, tarpons can breathe air from the surface. There are two species in a single genus *Megalops* in the family Megalopidae, one native to the Atlantic, and the other to the Indo-Pacific oceans.

The genus name derives from the Greek adjective *megalo* meaning 'large', and the noun *opsi*, meaning 'face'.

Telescopefish

Telescopefish are small, deep-sea aulopiform fish comprising the small family Giganturidae. There are just two known species, both within the genus *Gigatura*. Though rarely captured, they are found in cold, deep tropical to subtropical waters worldwide.

The common name of these fish are understandably named for their bizarre, tubular eyes. The genus name *Gigantura* is in reference to the Gigantes, a race of giants in Greek mythology—coupled with the suffix *oura*, meaning "tail", thus *Gigantura* refers to the ribbon-like lower half of the tailfin.

Physical Description

The Giganturidae are slender, slightly tapered fish with large heads dominated by large, forward-pointing telescoping eyes with large lenses. The head ends in a short, pointed snout. The highly extensile mouth is lined with sharp, slightly recurved and depressible teeth and it extends well past the eyes. The body lacks scales but is covered in easily abraded, silvery guanine which imparts a greenish to purplish iridescence in life. The gas bladder is absent and the stomach is highly distensible.

The transparent fins are spineless; the deeply forked and hypocercal caudal fin is most striking, with the lower lobe extended to a length exceeding that of the body. The pectoral fins are large, situated above the gill opening, and inserted horizontally. The anal fin and single dorsal fin are both situated far back of the head. The pelvic fins and adipose fin are absent.

Also absent are: the premaxilla, orbitosphenoid, parietal, symplectic, post-temporal, and supratemporal bones; the gill rakers; and the branchiostegal rays. The loss of these structures is attributed to neoteny; that is, the retention of larval characteristics.

Gigantura indica is the larger of the two species in terms of length at ca. 20.3 centimetres standard length (a measurement excluding the caudal fin). However, *Gigantura chuni* is slightly more robust in build.

Life History

Telescopefish are presumed to be solitary, active predators, frequenting the mesopelagic to bathypelagic zones of the water column, from ca. 500 metres down to ca. 3,000 metres. By using their tubular, large-lensed eyes—which are adapted for optimal binocular light collection, at the expense of lateral vision—telescopefish are likely able to spy their prey's weak bioluminescence from a distance, as well as (by looking skyward) resolve the outlined silhouettes of prey against the gloom above. Their eyes may also help telescopefish to better judge distance of prey; these visual adaptations are typical of deep-sea fish (cf. barrel-eye, tube-eye).

Common prey items include bristlemouths, lanternfish, and barbeled dragonfish. Owing to the telescopefishes' highly extensile jaws and distensible stomachs, they are able to swallow prey larger than themselves; this is also a common adaptation to life in the lean depths.

Much less is known of their reproductive habits. They are presumed to be non-guarding pelagic spawners, releasing eggs and sperm indiscriminately into the water. The fertilized eggs are buoyant and become incorporated into the zooplankton, wherein they and the larvae remain—likely at much shallower depths than the adults—until metamorphosis into juvenile or adult form.

Temperate Perch

The members of Percichthyidae family are known as the temperate perches. They belong to the Order Perciformes or perch-like fishes.

The name Percichthyidae derives from the Ancient Greek words *Perca* for 'perch' and *ichthyos* for 'fish' and translated basically means *perch-like fishes*.

The temperate perches are closely related to the temperate basses of the Moronidae Family, and older literature treats the latter as a belonging to the Percichthyidae Family. Australian freshwater Percichthyids were once placed in the Serranidae (marine grouper) family, and the two families are considered to have some link to each other.

Around 22 species of Percichthyids are now recognised, grouped in around 11 genera. Most but not all are freshwater fishes. They are mainly found in Australasia and South America, though some (notably genus *Siniperca,* which has the largest number of species) are native to Asia, and some of the marine species are found as far north in the Pacific as California.

Australia has the greatest number of Percichthyid species, where they are represented by the Australian freshwater cods (*Maccullochella spp.*), which are Murray cod, Mary River cod, eastern freshwater cod and trout cod, by the Australian freshwater blackfishes (*Gadopsis spp.*), which are river blackfish and two-spined blackfish, and by the Australian freshwater perches (*Macquaria spp.*), which are golden perch, Macquarie perch, Australian bass and estuary perch.

Several other Australian freshwater species also sit within the Percichthyidae Family while research using mitochondrial DNA suggests the species of the Nannopercidae family are in reality Percichthyids as well. Australia is unique in having a freshwater fish fauna dominated by Percichthyids and allied families/species. This in contrast to Europe and Asia whose fish faunas are dominated by members of the Cyprinidae carp family. (Indeed, Australia does not have a single naturally occurring Cyprinid species; unfortunately the illegal introduction of carp has now established the family's presence in Australia.)

A number of species are or have been important food species; some of these (e.g. the Murray cod, *Maccullochella peelii peelii*) have become threatened through overfishing and river regulation, while others (e.g. the Chinese perch, *Siniperca chuatsi*), are now farmed to some extent. Some smaller species (e.g. Balston's pygmy perch, *Nannatherina balstoni*) are popular in aquaria.

The extremely rare Bloomfield River cod, *Guyu wujalwujalensis*, is only found in a short stretch of the Bloomfield River in north Queensland.

Tench

The tench or doctor fish (*Tinca tinca*) is a freshwater and brackish water fish of the cyprinid family found throughout Eurasia from Western Europe including the British Isles east into Asia as far as the Ob and Yenisei Rivers. It is also found in Lake Baikal. It normally inhabits slow-moving freshwater habitats, particularly lakes and lowland rivers.

Ecology

The tench is most often found in still waters with a clayey or muddy substrate and abundant vegetation. This species is rare in clear waters across stony ground, and is absent altogether from fast-flowing streams. It tolerates water with a low oxygen concentration, even being found in waters where the carp cannot survive.

Tench feed mostly at night on algae and benthic invertbrates of various kinds that they root up from the bottom.

Breeding takes place in shallow water usually among aquatic plants where the sticky green eggs can be deposited. Spawning usually occurs in summer, and as many as three hundred thousand eggs may be produced. Growth is rapid, and fish may reach a weight of 0.11 kg (0.25 lb) within the first year.

Morphology

Tench have a stocky, carp-like shape, olive-green skin, darker above and almost golden below. The caudal fin is square in shape. The other fins are distinctly rounded in shape. The mouth is rather narrow and provided at each corner with a very small barbel. Maximum size is 70 cm, though most specimens are very much smaller. The eyes are small and red-orange in colour. Sexual dimorphism is weak, limited to the adult females having a more convex ventral profile when compared with males. Males may also possess a very thick and flattened outer ray to the ventral fins.

The tench has very small scales, which are deeply imbedded in a thick skin, making it as slippery as an eel. Folklore has it that this slime cured any sick fish that rubbed against it, and from this belief arose the name doctor fish.

Golden Tench

An artificially-bred variety of tench called the golden tench or schlei is a popular ornamental fish for ponds. This form varies in colour from pale gold through to red, and some fish have black or red spots on the flanks and fins. Though somewhat similar to the goldfish, because these fish have such small scales, the golden quality is rather different. This variety is said to have been originally bred in Silesia.

Economic Significance

Tench are edible and considered to have a fine flavour, working well in recipes that would otherwise call for carp. They are an important target for coarse anglers and are also used as fodder for predatory species such as bass. Tench, particularly golden tench, are also kept as ornamental fish in ponds and less frequently aquaria.

Angling

Large tench may be found in gravel pits or deep, slow-moving waters with a clayey or silty bottom and lots of aquatic vegetation. They take a variety of baits but are "nibblers" and are difficult to hook. Fish over 1 kg (2 lb) in weight are very strong fighters when caught on a rod.

Tetra

Tetra are species of small South American freshwater fish, belonging to the family Characidae and to its former subfamily Alestiidae (the "African tetras"). The characidae are distinguished from other fish by the presence of a small adipose fin between the dorsal fin and caudal fin. Many of these, such as the neon tetra, are brightly coloured and easy to keep in captivity, and extremely popular for home aquaria.

The term tetra is not actually a taxonomic, phylogenetic term. Because of the popularity of tetras in the fishkeeping hobby, many unrelated fish are commonly known as tetras, including species from differing families. Even fish that are vastly different may be called tetras, such as *Hydrolycus scomberoides*, occasionally known as the sabretooth tetra or vampire tetra.

Though the list below is sorted by common name, in a number of cases, the common name is applied to different species, depending on country and context. Since the aquarium trade may use a different name for the same species, advanced aquarists tend to use scientific names for the less-common tetras. The list below is incomplete.

Tetra species:

- Adonis tetra, *Lepidarchus adonis*
- African long-finned tetra, *Brycinus longipinnis*
- African moon tetra, *Bathyaethiops caudomaculatus*
- Arnold's tetra, *Arnoldichthys spilopterus*
- Buenos Aires tetra, *Hyphessobrycon anisitsi*
- Banded tetra, *Astyanax fasciatus*
- Bandtail tetra, *Moenkhausia dichroura*
- Barred glass tetra, *Phenagoniates macrolepis*
- Bass tetra, *Hemigrammus rodwayi*

- Beacon tetra, *Hemigrammus ocellifer*
- Belgian flag tetra, *Hyphessobrycon heterorhabdus*
- Black darter tetra, *Poecilocharax weitzmani*
- Black morpho tetra, *Poecilocharax weitzmani*
- Black neon tetra, *Hyphessobrycon herbertaxelrodi*
- Black phantom tetra, *Hyphessobrycon megalopterus*
- Black tetra, *Gymnocorymbus ternetzi*
- Black tetra, *Gymnocorymbus thayer*
- Black wedge tetra, *Hemigrammus pulcher*
- Blackband tetra, *Hyphessobrycon scholzei*
- Blackedge tetra, *Tyttocharax madeirae*
- Black-flag tetra, *Hyphessobrycon rosaceus*
- Black-jacket tetra, *Moenkhausia takasei*
- Blackline tetra, *Hyphessobrycon scholzei*
- Bleeding heart tetra, *Hyphessobrycon erythrostigma*
- Bleeding heart tetra, *Hyphessobrycon socolofi*
- Blind tetra, *Stygichthys typhlops*
- Bloodfin tetra, *Aphyocharax anisitsi*
- Bloodfin tetra, *Aphyocharax alburnus*
- Blue tetra, *Mimagoniates microlepis*
- Blue tetra, *Tyttocharax madeirae*
- Blue tetra, *Boehlkea fredcochui*
- Brilliant rummynose tetra, *Hemigrammus bleheri*
- Bucktooth tetra, *Exodon paradoxus*
- Buenos Aires tetra, *Hyphessobrycon anisitsi*
- Butterfly tetra, *Gymnocorymbus ternetzi*
- Callistus tetra, *Hyphessobrycon eques*
- Calypso tetra, *Hyphessobrycon axelrodi*
- Cardinal tetra, *Paracheirodon axelrodi*
- Carlana tetra, *Carlana eigenmanni*

- Cochu's blue tetra, *Boehlkea fredcochui*
- Central tetra, *Astyanax aeneus*
- Coffee-bean tetra, *Hyphessobrycon takasei*
- Colcibolca tetra, *Astyanax nasutus*
- Congo tetra, *Phenacogrammus interruptus*
- Copper tetra, *Hasemania melanura*
- Costello tetra, *Hemigrammus hyanuary*
- Creek tetra, *Bryconamericus scleroparius*
- Creek tetra, *Bryconamericus terrabensis*
- Croaking tetra, *Mimagoniates inequalis*
- Croaking tetra, *Mimagoniates lateralis*
- Dawn tetra, *Hyphessobrycon eos*
- Dawn tetra, *Aphyocharax paraguayensis*
- Diamond tetra, *Moenkhausia pittieri*
- Discus tetra, *Brachychalcinus orbicularis*
- Disk tetra, *Brachychalcinus orbicularis*
- Disk tetra, *Myleus schomburgkii*
- Dragonfin tetra, *Pseudocorynopoma doriae*
- Ember tetra, *Hyphessobrycon amandae*
- Emperor tetra, *Nematobrycon palmeri*
- False black tetra, *Gymnocorymbus thayeri*
- False neon tetra, *Paracheirodon simulans*
- False red nose tetra, *Petitella georgiae*
- False rummynose tetra, *Petitella georgiae*
- Featherfin tetra, *Hemigrammus unilineatus*
- Firehead tetra, *Hemigrammus bleheri*
- Flag tetra, *Hyphessobrycon heterorhabdus*
- Flame tail tetra, *Aphyocharax erythrurus*
- Flame tetra, *Hyphessobrycon flammeus*
- Garnet tetra, *Hemigrammus pulcher*

- Glass tetra, *Moenkhausia oligolepis*
- Glossy tetra, *Moenkhausia oligolepis*
- Glowlight tetra, *Hemigrammus erythrozonus*
- Gold tetra, *Hemigrammus rodwayi*
- Golden tetra, *Hemigrammus rodwayi*
- Goldencrown tetra, *Aphyocharax alburnus*
- Goldspotted tetra, *Hyphessobrycon griemi*
- Gold-tailed tetra, *Carlastyanax aurocaudatus*
- Green dwarf tetra, *Odontocharacidium aphanes*
- Green neon tetra, *Paracheirodon simulans*
- Green tetra, *Paracheirodon simulans*
- Griem's tetra, *Hyphessobrycon griemi*
- Head-and-taillight tetra, *Hemigrammus ocellifer*
- January tetra, *Hemigrammus hyanuary*
- Jellybean tetra, *Lepidarchus adonis*
- Jewel tetra, *Hyphessobrycon eques*
- Jumping tetra, *Hemibrycon tridens*
- Largespot tetra, *Astyanax orthodus*
- Lemon tetra, *Hyphessobrycon pulchripinnis*
- Longfin tetra, *Brycinus longipinnis*
- Long-finned glass tetra, *Xenagoniates bondi*
- Longjaw tetra, *Bramocharax bransfordii*
- Loreto tetra, *Hyphessobrycon loretoensis*
- Mayan tetra, *Hyphessobrycon compressus*
- Mexican tetra, *Astyanax mexicanus*
- Mimic scale-eating tetra, *Probolodus heterostomus*
- Mourning tetra, *Brycon pesu*
- Naked tetra, *Gymnocharacinus bergii*
- Neon tetra, *Paracheirodon innesi*
- Niger tetra, *Arnoldichthys spilopterus*

- Nurse tetra, *Brycinus nurse*
- Oneline tetra, *Nannaethiops unitaeniatus*
- One-line tetra, *Hemigrammus unilineatus*
- Orangefin tetra, *Bryconops affinis*
- Ornate tetra, *Hyphessobrycon bentosi*
- Panama tetra, *Hyphessobrycon panamensis*
- Penguin tetra, *Thayeria boehlkei*
- Peruvian tetra, *Hyphessobrycon peruvianus*
- Petticoat tetra, *Gymnocorymbus ternetzi*
- Phantom tetra, *Hyphessobrycon megalopterus*
- Pittier's tetra, *Moenkhausia pittieri*
- Pretty tetra, *Hemigrammus pulcher*
- Pristella tetra, *Pristella maxillaris*
- Pygmy tetra, *Odontostilbe dialeptura*
- Rainbow tetra, *Nematobrycon palmeri*
- Rainbow tetra, *Nematobrycon lacortei*
- Red eye tetra, *Moenkhausia sanctaefilomenae*
- Red phantom tetra, *Hyphessobrycon sweglesi*
- Red tetra, *Hyphessobrycon flammeus*
- Redeye tetra, *Moenkhausia sanctaefilomenae*
- Redeye tetra, *Moenkhausia oligolepis*
- Rednose tetra, *Hemigrammus rhodostomus*
- Red-nose tetra, *Hemigrammus bleheri*
- Redspotted tetra, *Copeina guttata*
- Rosy tetra, *Hyphessobrycon bentosi*
- Rosy tetra, *Hyphessobrycon rosaceus*
- Royal tetra, *Inpaichthys kerri*
- Ruby tetra, *Axelrodia riesei*
- Rummy-nose tetra, *Hemigrammus rhodostomus*
- Rummy-nose tetra, *Hemigrammus bleheri*

- Sailfin tetra, *Crenuchus spilurus*
- Savage tetra, *Hyphessobrycon savagei*
- Savanna tetra, *Hyphessobrycon stegemanni*
- Semaphore tetra, *Pterobrycon myrnae*
- Serpae tetra, *Hyphessobrycon eques*
- Sharptooth tetra, *Micralestes acutidens*
- Silver tetra, *Gymnocorymbus thayeri*
- Silver tetra, *Ctenobrycon spilurus*
- Silver tetra, *Micralestes acutidens*
- Silvertip tetra, *Hasemania melanura*
- Silvertip tetra, *Hasemania nana*
- Silver-tipped tetra, *Hasemania nana*
- Splash tetra, *Copella arnoldi*
- Splashing tetra, *Copella arnoldi*
- Spotfin tetra, *Hyphessobrycon socolofi*
- Spottail tetra, *Moenkhausia dichroura*
- Spotted tetra, *Copella nattereri*
- Swegles's tetra, *Hyphessobrycon sweglesi*
- Tailspot tetra, *Bryconops caudomaculatus*
- Tetra von Rio, *Hyphessobrycon flammeus*
- Three-lined African tetra, *Neolebias trilineatus*
- Tiete tetra, *Brycon insignis*
- Tortuguero tetra, *Hyphessobrycon tortuguerae*
- Transparent tetra, *Charax gibbosus*
- True big-scale tetra, *Brycinus macrolepidotus*
- Uruguay tetra, *Cheirodon interruptus*
- White spot tetra, *Aphyocharax paraguayensis*
- X-ray tetra, *Pristella maxillaris*
- Yellow tetra, *Hyphessobrycon bifasciatus*
- Yellow-tailed African tetra, *Alestopetersius caudalis*

Thornfish

The thornfishes are a family, Bovichtidae, of fishes in the order Perciformes. The family is spelled Bovichthyidae in J. S. Nelson's *Fishes of the World*.

They are native to coastal waters off Australia, New Zealand, and South America, and to rivers and lakes of southeast Australia and Tasmania.

Genera

There are ten species in three genera.

- Genus *Bovichtus* Valenciennes in Cuvier and Valenciennes, 1832
  - — *Bovichtus angustifrons* (Regan, 1913).
  - — *Bovichtus argentinus* (MacDonagh, 1931).
  - — *Bovichtus chilensis* (Regan, 1913).
  - — *Bovichtus diacanthus* (Carmichael, 1819).
  - — *Bovichtus oculus* (Hardy, 1989).
  - — *Bovichtus psychrolutes* (Gunther, 1860).
  - — Thornfish, *Bovichtus variegatus* (Richardson, 1846).
  - — *Bovichtus veneris* (Sauvage, 1879).
- Genus *Cottoperca* Steindachner, 1876
  - — *Cottoperca gobio* (Gunther, 1861).
- Genus *Halaphritis* (Last, Balushkin and Hutchins, 2002)
  - — *Halaphritis platycephala* (Last, Balushkin & Hutchins, 2002).

Threadfin

Threadfins are silvery grey perciform marine fish of the family Polynemidae. Found in tropical to subtropical waters throughout the world, the threadfin family contains nine genera and 33 species. An unrelated species sometimes known by the name threadfin, *Alectis indicus*, is properly known as the Indian threadfish (family Carangidae).

Ranging in length from 20 centimetres in the black-finned

threadfin (*Polydactylus nigripinnis*) to 200 centimetres in fourfinger threadfins (*Eleutheronema tetradactylum*) and giant African threadfins (*Polydactylus quadrifilis*), threadfins are both important to commercial fisheries as a food fish, and popular among anglers. Their habit of forming large schools makes the threadfins a reliable and economic catch.

Their bodies are elongate and fusiform, with spinous and soft dorsal fins widely separate. Their tail fins are large and deeply forked; this is an indication of their speed and agility. The mouth is large and inferior; a blunt snout projects far ahead. The jaws and palate possess bands of *villiform* (fibrous) teeth.

The most distinguishing feature of the threadfins are their pectoral fins: they are composed of two distinct sections, the lower of which consisting of between 3-7 long, thread-like independent rays. In *Polynemus* species there may be up to 15 of these modified rays.

In some species, such as the royal threadfin (*Pentanemus quinquarius*), the thread-like rays may extent well past the tail fin. This feature explains both the common name *threadfin* and the family name *Polynemidae,* from the Greek *poly* meaning "many" and *nema* meaning "filament." Similar species, such as the mullets (family Mugilidae) and milkfish (family Chanidae) can be easily distinguished from threadfins by their lack of filamentous pectoral rays.

Threadfins frequent open, shallow water in areas with muddy, sandy or silty bottoms; they are rarely glimpsed at reefs. Their filamentous pectoral rays are thought to serve as tactile structures, helping the fish to find prey within the sediments. Noted for being euryhaline, threadfins are able to tolerate a wide range of salinity levels. This attribute allows threadfins to enter estuaries and even rivers. They feed primarily on crustaceans and smaller fish.

Presumed to be pelagic spawners, threadfins probably release many tiny buoyant eggs into the water column which then become part of the plankton. The eggs float freely with the currents until hatching. In Hawaii, sixfinger threadfins (Polydactylus sexfilis) are the subject of experimental aquaculture; they may soon alleviate pressures on wild stocks and lessen the state's dependence on imported fish.

Species

There are about forty species in eight genera:

- Genus *Eleutheronema*
  - — East Asian fourfinger threadfin, *Eleutheronema rhadinum* (Jordan & Evermann, 1902).
  - — Fourfinger threadfin, *Eleutheronema tetradactylum* (Shaw, 1804).
  - — Threefinger threadfin, *Eleutheronema tridactylum* (Bleeker, 1849).
- Genus *Filimanus*
  - — Sevenfinger threadfin, *Filimanus heptadactyla* (Cuvier, 1829).
  - — Javanese threadfin, *Filimanus hexanema* (Cuvier, 1829).
  - — Splendid threadfin, *Filimanus perplexa* Feltes, 1991.
  - — Eightfinger threadfin, *Filimanus sealei* (Jordan & Richardson, 1910).
  - — *Filimanus similis* Feltes, 1991.
  - — Yellowthread threadfin, *Filimanus xanthonema* (Valenciennes, 1831).
- Genus *Galeoides*
  - — Lesser African threadfin, *Galeoides decadactylus* (Bloch, 1795).
- Genus *Leptomelanosoma*
  - — Indian threadfin, *Leptomelanosoma indicum* (Shaw, 1804).
- Genus *Parapolynemus*
  - — Dwarf paradise fish, *Parapolynemus verekeri* (Saville-Kent, 1889).
- Genus *Pentanemus*
  - — Royal threadfin, *Pentanemus quinquarius* (Linnaeus, 1758).

- Genus *Polydactylus*
  - — Blue bobo, *Polydactylus approximans* (Lay & Bennett, 1839).
  - — Slender fivefinger threadfin, *Polydactylus bifurcus* (Motomura, Kimura & Iwatsuki, 2001).
  - — Long-limb threadfin, *Polydactylus longipes* (Motomura, Okamoto & Iwatsuki, 2001).
  - — King threadfin, *Polydactylus macrochir* (Gunther, 1867).
  - — River threadfin, *Polydactylus macrophthalmus* (Bleeker, 1858).
  - — African blackspot threadfin, *Polydactylus malagasyensis* (Motomura & Iwatsuki, 2001).
  - — Small-mouthed threadfin, *Polydactylus microstomus* (Bleeker, 1851).
  - — *Polydactylus mullani* (Hora, 1926).
  - — Australian threadfin, *Polydactylus multiradiatus* (Gunther, 1860).
  - — Black-finned threadfin, *Polydactylus nigripinnis* (Munro, 1964).
  - — Atlantic threadfin, *Polydactylus octonemus* (Girard, 1858).
  - — Littlescale threadfin, *Polydactylus oligodon* (Gunther, 1860).
  - — Yellow bobo, *Polydactylus opercularis* (Seale & Bean, 1907).
  - — Persian blackspot threadfin, *Polydactylus persicus* (Motomura & Iwatsuki, 2001).
  - — Striped threadfin, *Polydactylus plebeius* (Broussonet, 1782).
  - — Giant African threadfin, *Polydactylus quadrifilis* (Cuvier, 1829).
  - — Sixfinger threadfin, *Polydactylus sexfilis* (Valenciennes, 1831).
  - — Blackspot threadfin, *Polydactylus sextarius* (Bloch & Schneider, 1801).
  - — *Polydactylus siamensis* (Motomura, Iwatsuki & Yoshino, 2001).
  - — Barbu, *Polydactylus virginicus* (Linnaeus, 1758).

- Genus *Polynemus*
  - — Northern paradise fish, *Polynemus aquilonaris* (Motomura, 2003).
  - — *Polynemus bidentatus* (Motomura & Tsukawaki, 2006).
  - — Eastern paradise fish, *Polynemus dubius* (Bleeker, 1854).
  - — Hornaday's paradise fish, *Polynemus hornadayi* (Myers, 1936).
  - — Elegant paradiseus fish, *Polynemus multifilis* (Temminck & Schlegel, 1843).
  - — Paradise threadfin, *Polynemus paradiseus* (Linnaeus, 1758).

Threadfin Bream

The threadfin breams are a family, Nemipteridae, of fishes in the order Perciformes. They are also known as whiptail breams and false snappers.

They are found in tropical waters of the Indian and western Pacific Oceans. Most species are benthic (bottom-feeding) carnivorous, eating small fishes, cephalopods, crustaceans and polychaetes; however, a few species eat plankton.

Species

There are about sixty species in five genera:

- Genus *Nemipterus* (Swainson, 1839).
  - — Yellow-lip threadfin bream, *Nemipterus aurifilum* (Ogilby, 1910).
  - — Dawn threadfin bream, *Nemipterus aurorus* (Russell, 1993).
  - — Balinese threadfin bream, *Nemipterus balinensis* (Bleeker, 1859).
  - — Dwarf threadfin bream, *Nemipterus balinensoides* (Popta, 1918).
  - — Yellowbelly threadfin bream, *Nemipterus bathybius* (Snyder, 1911).
  - — Delagoa threadfin bream, *Nemipterus bipunctatus* (Valenciennes, 1830).

— Celebes threadfin bream, *Nemipterus celebicus* (Bleeker, 1854).

— Fork-tailed threadfin bream, *Nemipterus furcosus* (Valenciennes, 1830).

— Graceful threadfin bream, *Nemipterus gracilis* (Bleeker, 1873).

— Ornate threadfin bream, *Nemipterus hexodon* (Quoy & Gaimard, 1824).

— Teardrop threadfin bream, *Nemipterus isacanthus* (Bleeker, 1873).

— Japanese threadfin bream, *Nemipterus japonicus* (Bloch, 1793).

— Red filament threadfin bream, *Nemipterus marginatus* (Valenciennes, 1830).

— Mauvelip threadfin bream, *Nemipterus mesoprion* (Bleeker, 1853).

— Doublewhip threadfin bream, *Nemipterus nematophorus* (Bleeker, 1853).

— Yellow-tipped threadfin bream, *Nemipterus nematopus* (Bleeker, 1851).

— Redspine threadfin bream, *Nemipterus nemurus* (Bleeker, 1857).

— Notchedfin threadfin bream, *Nemipterus peronii* (Valenciennes, 1830).

— Randall's threadfin bream, *Nemipterus randalli* (Russell, 1986).

— Fivelined threadfin bream, *Nemipterus tambuloides* (Bleeker, 1853).

— Theodore's threadfin bream, *Nemipterus theodorei* (Ogilby, 1916).

— Palefin threadfin bream, *Nemipterus thosaporni* (Russell, 1991).

— Golden threadfin bream, *Nemipterus virgatus* (Houttuyn, 1782).

— Fiji threadfin bream, *Nemipterus vitiensis* (Russell, 1990).

— Slender threadfin bream, *Nemipterus zysron* (Bleeker, 1857).

- Genus *Parascolopsis* Boulenger, 1901

— Smooth dwarf monocle bream, *Parascolopsis aspinosa* (Rao & Rao, 1981).

— *Parascolopsis baranesi* (Russell & Golani, 1993).

— Redfin dwarf monocle bream, *Parascolopsis boesemani* (Rao & Rao, 1981).

— *Parascolopsis capitinis* (Russell, 1996).

— Rosy dwarf monocle bream, *Parascolopsis eriomma* (Jordan & Richardson, 1909).

— Unarmed dwarf monocle bream, *Parascolopsis inermis* (Temminck & Schlegel, 1843).

— Dwarf monocle bream, *Parascolopsis melanophrys* (Russell & Chin, 1996).

— Slender dwarf monocle bream, *Parascolopsis qantasi* (Russell & Gloerfelt-Tarp, 1984).

— Red-spot dwarf monocle bream, *Parascolopsis rufomaculatus* (Russell, 1986).

— Long-rayed dwarf monocle bream, *Parascolopsis tanyactis* (Russell, 1986).

— Tosa dwarf monocle bream, *Parascolopsis tosensis* (Kamohara, 1938).

— Scaly dwarf monocle bream, *Parascolopsis townsendi* (Boulenger, 1901).

- Genus *Pentapodus* (Quoy and Gaimard, 1824).

— *Pentapodus aureofasciatus* (Russell, 2001).

— White-shouldered whiptail, *Pentapodus bifasciatus* (Bleeker, 1848).

— Small-toothed whiptail, *Pentapodus caninus* (Cuvier, 1830).

— Double whiptail, *Pentapodus emeryii* (Richardson, 1843).
— Japanese whiptail, *Pentapodus nagasakiensis* (Tanaka, 1915).
— Paradise whiptail, *Pentapodus paradiseus* (Gunther, 1859).
— Northwest Australian whiptail, *Pentapodus porosus* (Valenciennes, 1830).
— Butterfly whiptail, *Pentapodus setosus* (Valenciennes, 1830).
— Three-striped whiptail, *Pentapodus trivittatus* (Bloch, 1791).
— Striped whiptail, *Pentapodus vitta* (Quoy & Gaimard, 1824).

- Genus *Scaevius* (Whitley, 1947).

— Green-striped coral bream, *Scaevius milii* (Bory de Saint-Vincent, 1823).

- Genus *Scolopsis* Cuvier, 1814

— Peters' monocle bream, *Scolopsis affinis* (Peters, 1877).
— Yellowstripe monocle bream, *Scolopsis aurata* (Park, 1797).
— Two-lined monocle bream, *Scolopsis bilineata* (Bloch, 1793).
— Thumbprint monocle bream, *Scolopsis bimaculatus* (Ruppell, 1828).
— Saw-jawed monocle bream, *Scolopsis ciliata* (Lacepede, 1802).
— Bridled monocle bream, *Scolopsis frenatus* (Gunther, 1859).
— Arabian monocle bream, *Scolopsis ghanam* (Forsskal, 1775).
— Striped monocle bream, *Scolopsis lineata* (Quoy & Gaimard, 1824).
— Pearly monocle bream, *Scolopsis margaritifera* (Cuvier, 1830).

— Monogrammed monocle bream, *Scolopsis monogramma* (Cuvier, 1830).

— Black-streaked monocle bream, *Scolopsis taeniatus* (Cuvier, 1830).

— Lattice monocle bream, *Scolopsis taenioptera* (Valenciennes, 1830).

— Bald-spot monocle bream, *Scolopsis temporalis* (Cuvier, 1830).

— Three-lined monocle bream, *Scolopsis trilineata* (Kner, 1868).

— Whitecheek monocle bream, *Scolopsis vosmeri* (Bloch, 1792).

— Oblique-barred monocle bream, *Scolopsis xenochrous* (Gunther, 1872).

Three Spot Gourami

The three spot gourami, *Trichogaster trichopterus*, is a labyrinth fish. The natural colouration of this fish gets its name from the two spots along each side of its body in line with the eye, considered the third spot.

Marketed fresh or salted and pickled, they are also popular in the fishkeeping hobby, commonly kept in aquariums. Depending on the various artificial colour morphs, the species may be known under other common names including the Blue gourami, Cosby gourami, Gold gourami, Opaline gourami, and Two-spot gouramy.

Distribution and Ecology

Three spot gourami are endemic to the Mekong basin in Cambodia, Laos, Thailand and Vietnam and Yunnan in South East Asia. These fish live in marshes, swamps, canals, and lowland wetlands. They migrate during the flood season from permanent water bodies to flooded areas, such as seasonally flooded forests in the middle and lower Mekong. During the dry season, they will return to these permanent water bodies. These fish feed on zooplankton, crustaceans and insect larvae.

The male builds a bubble nest for the eggs, which he protects aggressively.

In the Aquarium

The Three spot gourami is a hardy fish. This species can be housed with a variety of tank mates that are of similar size and temperament. While males can be territorial with each other, they become timid around other, more aggressive fish. The ideal tank setup would be an aquarium of a minimum of 20 gallons and have plenty of live plants as well as rocks and driftwood for use as hiding places.

Feeding

The Three spot gourami is an omnivore and requires both algae-based foods as well as meaty foods. An algae-based flake food, along with freeze-dried bloodworms, tubifex worms, and brine shrimp will provide these fish with the proper nutrition. The average weight of a blue gourami is 2-3.5 ounces.

Breeding

The best way to differentiate between the male and female Three spot gourami is by the dorsal fin. In the male, the dorsal fin is long and pointed, while the female's is shorter and rounded. For breeding, the temperature should be about 26°C (78°F). When ready to breed, the male builds a bubble nest and then begins to entice the female by swimming back and forth, flaring his fins and raising his tail.

When this behaviour is noticed, the water level should be reduced to 6 inches. The depth is not extremely important as it is in other labyrinth fish. The female may lay up to 700 to 800 eggs. After spawning the female should be removed to a separate aquarium as the male may become aggressive towards her. The male protects the eggs and fry, but should be removed after they become free-swimming.

After hatching, there should be frequent water changes, especially during the third week, as this is when the labyrinth organ is developing. The fry should be fed infusoria and nauplii.

Three-spined Stickleback

Distribution and Morphological Variation: The three spined stickleback, *Gasterosteus aculeatus*, is a fish native to much of northern Europe, northern Asia and North America. It has been introduced into parts of southern and central Europe.

Three subspecies are currently recognised by the IUCN:

- *Gasterosteus aculeatus aculeatus* is found in most of the species range, and is the subspecies most strictly termed the three-spined stickleback; its common name in England is the tiddler, although "tittlebat" is also sometimes used.
- *G. a. williamsoni*, the unarmoured threespine stickleback, is found only in North America; its recognised range is southern California, though there are isolated reports of it occurring in British Columbia and Mexico.
- *G. a. santaeannae*, the Santa Ana stickleback, is also restricted to North America.

These subspecies actually represent three examples from the enormous range of morphological variation present within three-spined sticklebacks. These fall into two rough categories, the anadromous and the freshwater forms.

The anadramous form spends most of its adult life eating plankton in the sea, and returns to freshwater to breed. The adult fish are typically between 6 and 10 cm long, and have 30 to 40 lateral armour plates along their sides.

They also have long dorsal and pelvic spines. The anadromous form is morphologically similar all around the Northern Hemisphere, such that anadromous fish from the Baltic, the Atlantic and the Pacific all resemble each other quite closely.

Three-spined stickleback populations are also found in freshwater lakes and streams. These populations were probably formed when anadromous fish started spending their entire life cycle in freshwater, and thus evolved to live there all year round. Freshwater populations are extremely morphologically diverse, to the extent that many observers (and some taxonomists) would describe a new subspecies of three-spined stickleback in almost every lake in the Northern Hemisphere. One consistent difference between freshwater populations and their anadromous ancestors is the amount of body armour, as the majority of freshwater fish only have between 0 and 12 lateral armour plates, and shorter dorsal and pelvic spines.

However, there are also large morphological differences between lakes. One major axis of variation is between populations found in deep, steep sided lakes and those in small, shallow lakes. The fish in the deep lakes typically feed in the surface waters on plankton, and often have large eyes, with short, slim bodies and an upturned jaw. Some researchers refer to this as the *limnetic* form.

Fish from shallow lakes feed mainly on the lake bed, and are often long and heavy bodied with a relatively horizontal jaw and a small eye. These populations are referred to as the *benthic* form.

Since each watershed was probably colonised separately by anadromous sticklebacks, it is widely believed that morphologically similar populations in different watersheds or on different continents evolved independently. There is a unique population in the meromictic Pink Lake in Gatineau Park, Quebec.

One fascinating aspect of this morphological variation is that a number of lakes contain both a limnetic and a benthic type, and these do not interbreed with each other. Evolutionary biologists often define species as populations that do not interbreed with each other (the Biological Species Concept), and thus the benthics and limnetics within each lake would constitute separate species. These species pairs are an excellent example of how adaptation to different environments (in this case feeding in the surface waters or on the lake bed) can generate new species. This process has come to be termed ecological speciation. This type of species pairs is found in British Columbia in Western Canada.

The lakes themselves only contain three-spined sticklebacks and cutthroat trout, and all are on islands. Tragically, the pair in Hadley Lake on Lasqueti Island was destroyed in the mid 1980s by the introduction of a predatory catfish, and the pair in Enos Lake on Vancouver Island has started to interbreed and are no longer two distinct species. The two remaining pairs are on Texada Island, in Paxton Lake and Priest Lake, and they are listed as Endangered in the Canadian Species At Risk Act.

Other species pairs which consist of a well-armoured marine form and a smaller, unarmoured fresh water form are being studied in ponds and lakes in Southcentral Alaska that were once marine habitats such as those uplifted during the 1964 Alaskan Earthquake.

The evolutionary dynamics of these species pairs are providing a model for the processes of speciation which has taken place in less than 20 years in at least one lake.

In 1982, a chemical eradication programme at Loberg Lake, Alaska, killed the resident freshwater populations of sticklebacks. Oceanic sticklebacks were introduced and colonised the lake. In just 12 years beginning in 1990, the frequency of the oceanic form dropped steadily, from 100 per cent to 11 per cent, while a variety with fewer plates increased to 75 per cent of the population, with various intermediate forms making up another small fraction (Carroll 2006).

This rapid evolution is thought to be possible through genetic variations that confer competitive advantages for survival in freshwater when conditions shift rapidly from salt to freshwater. However, the actual molecular basis of this evolution still remains unknown.

Although sticklebacks are found all round the coast of the Northern Hemisphere and are thus viewed by the IUCN as species of least concern, the unique evolutionary history encapsulated in many freshwater populations means they deserve more legal protection than they currently receive.

Life History

Many populations take 2 years to mature and experience only one breeding season before dying and some can take up to 3 years to reach maturity. However, some freshwater populations and populations at extreme latitudes can reach maturity in only 1 year.

In spring, males defend territories where they build nests on the bottom of the pond or other body of water; the sequence of territorial, courtship and mating behaviours was described in detail by Niko Tinbergen in a landmark early study in ethology.

Territorial males develop a red chin and belly colouration, and Tinbergen showed that the red colour acted as a simple sign stimulus, releasing aggression in other males and the first steps in the courtship sequence from gravid females. Red colouration is produced from carotenoids found in the diet of the fish. As carotenoids cannot be synthesised de novo, the degree of colouration gives an indication of male quality, with higher quality males showing more intense colouration.

However, it is noteworthy that the response to red is not universal across the entire species complex, with black throated populations often found in peat-stained waters. Males also develop blue irises on maturation. Only the males care for the eggs once they are fertilized. Parental care is intense, involving nest maintenance and fanning of the eggs to ensure a fresh water supply.

Males build the nests from vegetation, sand, pebbles and other debris, adhering the nest together with spiggin, a proteinaceous glue-like substance secreted from the kidneys. Sticklebacks have four colour photoreceptors in their retina, making them tetrachromatic. They are capable of perceiving ultraviolet wavelengths of light invisible to the human eye and use such wavelengths in their normal behavioural repertoire.

Genetics

Three-spined sticklebacks have recently become a major

research orgasm for evolutionary biologists trying to understand the genetic changes involved in adapting to new environments. The entire genome of a female fish from Bear Paw Lake in Alaska was recently sequenced by the Broad Institute and many other genetic resources are available. This population is under risk by the presence of introduced northern pike in a nearby lake.

Threetooth Puffer

The threetooth puffer, *Triodon macropterus,* is a tetraodontiform fish, the only species in the genus *Triodon* and family Triodontidae.

It is native to the Indian and western Pacific Oceans. Its name comes from the Greek *tria* meaning "three" and *odous* meaning "tooth", and refers to the three fused teeth making up a beak-like structure.

The threetooth puffer has a distinctive shape, with a huge belly flap as large as or larger than its body; it inflates this with seawater when threatened.

Thresher Shark

Thresher sharks are large lamniform sharks of the family Alopiidae. Found in all temperate and tropical oceans of the world, the family contains three species all within the genus *Alopias*.

Naming

The genus and family name derive from the Greek word *alopex,* meaning fox. Indeed the long-tailed thresher shark, *Alopias vulpinus,* is named the fox shark by some authorities.

Distribution and Habitat

Although occasionally sighted in shallow, inshore waters, thresher sharks are primarily pelagic; they prefer the open ocean, staying within the first 500 m of the water column. Common threshers tend to be more common in coastal waters over continental shelves. In the North Pacific, common thresher sharks are found along the continental shelves of North America and Asia. They are rare in the Central and Western Pacific. In the warmer waters of the Central & Western Pacific, bigeye and pelagic thresher sharks are more common.

Anatomy and Appearance

Named for and easily recognised by their exceptionally long, thresher-like tail or *caudal fins* (which account for 1/3 (33%) of their total body length), thresher sharks are active predators; the tail is actually used as a weapon to stun prey. By far the largest of the three species is the Common thresher, *Alopias vulpinus*, which may reach a length of 7.6 m (25 ft) and a weight of 348 kg (767 lb). The Bigeye thresher, *Alopias superciliosus*, is next in size, reaching a length of 4.9 m (16 ft); at just 3 m (10 ft), the Pelagic thresher, *Alopias pelagicus*, is the smallest.

Thresher sharks are fairly slender, with small dorsal fins and large, recurved pectoral fins. With the exception of the Bigeye thresher, these sharks have relatively small eyes. Colouration ranges from brownish, bluish or purplish gray dorsally with lighter shades ventrally. The three species can be roughly distinguished by the main colour of the dorsal surface of the body. Common threshers are dark green, Bigeye threshers are brown and Pelagic threshers are generally blue. Lighting conditions and water clarity can affect how any one shark appears to an observer, but the colour test is generally supported when other features are examined.

Diet

Pelagic schooling fish (such as bluefish, juvenile tuna, and mackerel), squid and cuttlefish are the primary food items of the thresher sharks. They are known to follow large schools of fish into shallow waters. Crustaceans and the odd seabird are also taken.

Behaviour

Thresher sharks are solitary creatures which keep to themselves. It is known that thresher populations of the Indian

Ocean are separated by depth and space according to gender. All species are noted for their highly migratory or oceanodromous habits.

Reproduction

No distinct breeding season is observed by thresher sharks. Fertilization and embryonic development occur internally; this ovoviviparous or live-bearing mode of reproduction results in a small litter (usually 2 to 4) of large well-developed pups, up to 150 cm at birth in thintail threshers. The young fish exhaust their yolk sacs while still inside the mother, at which time they begin feeding on the mother's unfertilized eggs; this is known as oophagy.

Thresher sharks are slow to mature, males reaching sexual maturity between 7 and 13 years of age and females between 8 and 14 years in bigeye threshers. They may live for 20 years or more.

Thresher sharks are one of the few shark species known to jump fully out of the water making turns like dolphins, this behaviour is called breaching.

Conservation

All three thresher shark species have been recently listed as vulnerable to extinction by the World Conservation Union (IUCN).

Importance to Humans

Like all large sharks, threshers are slow growing and are therefore threatened by commercial fisheries. Other than for its meat, the sharks are hunted for their liver oil, skin (for leather), and their fins, for use in shark-fin soup.

They do not appear to be a threat to humans, although some divers have been hit with the upper tail lobe. There is an unconfirmed account of a fisherman being decapitated by a tail swipe as the shark breached.

Thresher sharks are classified as prized gamefish in the United States and South Africa. Common thresher sharks are the target of a popular recreational fishery off Baja Mexico. Thresher sharks are managed in some areas for their value as both a recreational sport fish and commercial species.

Species

- Genus Alopias
  - — Pelagic thresher, *Alopias pelagicus*
  - — Bigeye thresher, *Alopias superciliosus*
  - — Common thresher, *Alopias vulpinus*

Tidewater Goby

The tidewater goby *Eucyclogobius newberryi* is a goby (Gobiidae) native to lagoons of streams along the coast of California. It is the sole member of its genus.

A small fish, only rarely longer than 5 cm (2 in), it is elongate with a blunt tail. Colour is a mottled grey, brown, or olive; living fish are somewhat translucent. The upper part of the first dorsal fin is clear or cream-coloured, while the second dorsal is longer than the first, and close in size to the anal fin. The large mouth extends back past the posterior edge of the eye, and is angled upwards. Unusually among gobies, the scales are cycloid instead of ctenoid; they are always absent from the head, and often from the underside too.

Despite the common name, this goby inhabits lagoons formed by streams running into the sea. The lagoons are blocked from the Pacific Ocean by sand bars, admitting salt water only during particular seasons, and so their water is brackish and cool. The tidewater goby prefers salinities of less than 10 ppt, and is thus more often found in the upper parts of the lagoons, near their inflow. (Juveniles have been found as far upstream as 12 km.) They also prefer sandy bottoms with depths of 20-100 cm, near emergent vegetation beds, since they breed in the open areas and winter over in the vegetation.

Their range extends from Tillas Slough at the mouth of the Smith River in Del Norte County, California, south to Agua Hedionda Lagoon in San Diego County. While once recorded in at least 87 coastal locations, they are now gone from many, including San Francisco Bay.

The tidewater goby was listed by the state of California for protection in 1987, and federally listed in 1994. However, there has been some controversy over this, since many populations in its range are apparently secure, and the fish is even abundant at times. However, the fish's need for specific kind of habitat means that the populations are isolated from each other, and subject to extirpation due to various human activities, such as draining of wetlands, sand bar breaches for the purpose of tidal flushing, pollutant accumulation in lagoons, and so forth. Even so, studies have shown that it is a resilient species, and populations have been successfully restored to wetlands that have been protected.

Tiger Barb

The tiger barb (*Puntius tetrazona*) or sumatra barb, is a species of tropical freshwater fish belonging to the Puntius genus of the minnow family. The natural geographic range reportedly extends throughout the Malay peninsula, Sumatra and Borneo, with unsubstantiated sightings reported in Cambodia. Tiger barbs are also found in many other parts of Asia, and with little reliable collection data over long periods of time, definite conclusions about their natural geographic range versus established introductions are difficult.

Physical Description

The tiger barb can grow to about 7 centimetres long and 3 centimetres wide, although they are often smaller when kept in captivity. Native fish are silver to brownish yellow with four vertical black stripes and red fins and snout.

Habitat

It has been reported that the tiger barb was found in clear or turbid shallow waters of moderately flowing streams. It lives in a tropical climate and prefers water with a 6.0-8.0 *pH,* a water hardness of 5-19 dGH, and a temperature range of 68-79 °F (20-26 °C). Its discovery in swamp lakes that are subject to great changes in water level suggests a wide tolerance to water quality fluctuations. Its average lifespan is 6 years.

Importance to Humans

The tiger barb is one of over 70 species of barb with commercial importance in the aquarium trade. Of the total ornamental fish species imported into the United States in 1992, only 20 species account for more than 60 per cent of the total number of species reported, with tiger barbs falling at tenth on the list with 2.6 million individuals imported. (Chapman *et al.* 1994).

Barbs that have been hybridised to emphasise bright colour combinations have grown in popularity and production over the last 20 years. Example hybrids of tiger barb include highly melanistic green tiger barbs that reflect green over their black because of the Tyndall effect, gold tiger barbs.

Name Origins

The current taxonomic status of the species is far from being settled. There has been debate over the years as to the appropriate genus and species for this fish. In 1855, the German ichthyologist Pieter Bleeker described this fish as *Barbus tetrazona*. In 1857, Bleeker described another species under the same name.

Then, in 1860, Bleeker used C. *sumatraus* to describe the original species. In the late 1930's, the mistake was discovered, and the tiger barb nomenclature was changed back to *B. tetrazona* (Alfred, 1963). More recently, Dr. L.P. Schultz has reclassified the barbs according to the number of barbers each species possesses (Axelrod and Sweeney, 1992).

However, as stated by some Zakaria-Ismail (1993), "from my ongoing osteological studies that have been classified under *Puntius*, the genus *Barbodes* cannot be properly defined." Today, we are left with three generic classifications, *Barbodes*, *Capoeta*, and *Puntius*, all of which appear in the literature when referring to tiger barbs and other barb species.

In the Aquarium

The tiger barb is an active schooling fish that is usually kept in groups of five or more. They are often aggressive in numbers less than 5 and are known fin nippers. If you only keep two in a tank, one will eventually chase the other. Semi-aggressive fish, they form a pecking order in the pack which they may extend to other fish, giving them a reputation for nipping at the fins of other fish, especially if they are wounded or injured. They are thus not recommended for tanks with slower, more peaceful fishes such as bettas, gouramis, angelfish and others with long flowing fins. They do however work well with many fast moving fish such as danios, playts and most catfish. When in large enough groups, however, they tend to spend most of their time chasing each other and leave other species of fish alone. They dwell primarily at the water's mid-level. Tiger barbs do best in soft, slightly acidic water. The tank should be well-lit with ample vegetation, about two-thirds of the tank space. These barbs are omnivorous and will consume processed foods such as flakes and crisps as well as live foods.

Breeding

The tiger barb usually attains sexual maturity at a body length of 20 to 30 millimetres (0.8 to 1.2 inches) in total length, or at approximately six to seven weeks of age. The females are larger with a rounder belly and a mainly black dorsal fin while the males have a bright, red nose with a distinct red line above the black on their dorsal fin. The egg-layers tend to spawn several hundred eggs in the early morning in clumps of plants. On average, 300 eggs can be expected from each spawn in a mature broodstock population, although the number of eggs released will increase with the maturity and size of the fish. Spawned eggs are adhesive, negatively buoyant in freshwater and average 1.18 ± 0.05 mm in diameter.

Tiger barbs have been documented to spawn as many as 500 eggs per female (Scheurmann 1990; Axelrod 1992). With proper conditioning, females can spawn at approximately two weeks intervals (Munro *et al.* 1990).

Once spawning is finished, they will usually eat any of the eggs that they find. It is usually necessary to separate the fish from the eggs after spawning in order to prevent the eggs from being eaten.

Common Hybrids

Inter- and intra-specific hybridisation is done to achieve different colours and patterns to satisfy market demand for new "tiger barb" varieties. Gold tiger barbs and albino tiger barbs are examples of commercially produced hybrid fish. Other examples are summarised below:

Common barb hybrids representing different colour patterns (Modified from Kortmulder 1972)

Female parent Species	*Common name*	*x*	*Male parent Species*	*Common name*
Puntius conchonius	rosy barb	x	*Puntius stoliczkanus*	tic-tac-toe barb
Puntius cumingii	cuming's barb	x	*Puntius stoliczkanus*	tic-tac-toe barb
Puntius stoliczkanus	tic-tac-toe barb	x	*Puntius cumingii*	cuming's barb
Puntius nigrofasciatus	black ruby barb	x	*Puntius stoliczkanus*	tic-tac-toe barb
Puntius stoliczkanus	tic-tac-toe barb	x	*Puntius nigrofasciatus*	black ruby barb
Puntius cumingii	cuming's barb	x	*Puntius nigrofasciatus*	black ruby barb
Puntius nigrofasciatus	black ruby barb	x	*Puntius conchonius*	rosy barb
Puntius tetrazona	tiger barb	x	*Puntius nigrofasciatus*	black ruby barb
Puntius conchonius	rosy barb	x	*Puntius tetrazona*	tiger barb
Puntius stoliczkanus	tic-tac-toe barb	x	*Puntius tetrazona*	tiger barb

3

Grunter and Triplespines

Terapontidae

Grunters or tigerperches are fishes in the family Terapontidae (also spelled Teraponidae, Theraponidae or Therapontidae). They are found in shallow coastal waters in the Indian Ocean and western Pacific, where they live in saltwater, brackish and freshwater habitats. They grow up to 80 cm in length and feed on fishes, insects and other invertebrates.

When caught, grunters make the characteristic grunting sounds that give them their name.

Species

There are 50 species in 15 genera:

- Genus *Amniataba*
  - — Tiger grunter, *Amniataba affinis* (Mees & Kailola, 1977).
  - — Yellowtail trumpeter, *Amniataba caudavittata* (Richardson, 1845).
  - — Barred grunter, *Amniataba percoides* (Gunther, 1864).
- Genus *Bidyanus*
  - — Bidyan perch, *Bidyanus bidyanus* (Mitchell, 1838).

- Genus *Hannia*
  - — Greenway's grunter, *Hannia greenwayi* (Vari, 1978).
- Genus *Hephaestus*
  - — Adamson's grunter, *Hephaestus adamsoni* (Trewavas, 1940).
  - — *Hephaestus carbo* (Ogilby & McCulloch, 1916).
  - — Long-nose sooty grunter, *Hephaestus epirrhinos* (Vari & Hutchins, 1978).
  - — Sooty grunter, *Hephaestus fuliginosus* (Macleay, 1883).
  - — Mountain grunter, *Hephaestus habbemai* (Weber, 1910).
  - — *Hephaestus komaensis* (Allen & Jebb, 1993).
  - — Lined grunter, *Hephaestus lineatus* (Allen, 1984).
  - — *Hephaestus obtusifrons* (Mees & Kailola, 1977).
  - — Raymond's grunter, *Hephaestus raymondi* (Mees & Kailola, 1977).
  - — Roemer's grunter, *Hephaestus roemeri* (Weber, 1910).
  - — Sepik grunter, *Hephaestus transmontanus* (Mees & Kailola, 1977).
  - — Threespot grunter, *Hephaestus trimaculatus* (Macleay, 1883).
- Genus *Lagusia*
  - — *Lagusia micracanthus* (Bleeker, 1860).
- Genus *Leiopotherapon*
  - — Fortescue grunter, *Leiopotherapon aheneus* (Mees, 1963).
  - — Large-scale grunter, *Leiopotherapon macrolepis* (Vari, 1978).
  - — *Leiopotherapon plumbeus* (Kner, 1864).
  - — Spangled perch, *Leiopotherapon unicolor* (Gunther, 1859).
- Genus *Mesopristes*
  - — Silver grunter, *Mesopristes argenteus* (De vis, 1884).
  - — Tapiroid grunter, *Mesopristes cancellatus* (Cuvier, 1829).
  - — Plain terapon, *Mesopristes elongatus* (Guichenot, 1866).
  - — *Mesopristes iravi* (Yoshino, Yoshigou & Senou, 2002).

- Genus *Pelates*
  - — Western striped trumpeter, *Pelates octolineatus* (Jenyns, 1840).
  - — Fourlined terapon, *Pelates quadrilineatus* (Bloch, 1790).
  - — Six-lined trumpeter, *Pelates sexlineatus* (Quoy & Gaimard, 1825).
- Genus *Pelsartia*
  - — *Pelsartia humeralis* (Ogilby, 1899).
- Genus *Pingalla*
  - — Gilbert's grunter, *Pingalla gilberti* (Whitley, 1955).
  - — Lorentz's grunter, *Pingalla lorentzi* (Weber, 1910).
  - — Black-blotch grunter, *Pingalla midgleyi* (Allen & Merrick, 1984).
- Genus *Rhynchopelates*
  - — *Rhynchopelates oxyrhynchus* (Temminck & Schlegel, 1842).
- Genus *Scortum*
  - — Leathery grunter, *Scortum hillii* (Castelnau, 1878).
  - — Small-headed grunter, *Scortum parviceps* (Macleay, 1883).
- Genus *Syncomistes*
  - — Kimberley grunter, *Syncomistes kimberleyensis* (Vari, 1978).
  - — Drysdale grunter, *Syncomistes rastellus* (Vari & Hutchins, 1978).
- Genus *Terapon*
  - — Jarbua terapon, *Terapon jarbua* (Forsskal, 1775).
  - — Small-scaled terapon, *Terapon puta* (Cuvier, 1829).
  - — Largescaled therapon, *Terapon theraps* (Cuvier, 1829).
- Genus *Variichthys*
  - — Jamur Lake grunter, *Variichthys jamoerensis* (Mees, 1971).
  - — Lake grunter, *Variichthys lacustris* (Mees & Kailola, 1977).

Tiger Shark

The tiger shark, *Galeocerdo cuvier*, one of the largest sharks, is the only member of the genus Galeocerdo. Mature sharks average 3.25 metres (11 ft) to 4.25 metres (14 ft) (10 to 14 ft) and weigh 385 to 635 kg (850 to 1400 lb). It is found in many of the tropical and temperate regions of the world's oceans, and is especially common around islands in the central Pacific. This shark is a solitary hunter, usually hunting at night. Its name is derived from the dark stripes down its body, which fade as the shark matures.

The tiger shark is a dangerous predator, known for eating a wide range of items. Its usual diet consists of fish, seals, birds, smaller sharks, squid, and turtles. It has sometimes been found with man-made waste such as license plates or pieces of old tires in its digestive tract. It is notorious for attacks on swimmers, divers and surfers in Hawaii; and is often referred to as the "bane of Hawaiian surfers" and "the wastebasket of the sea".

The tiger shark is second only to the great white in number of recorded human fatalities and is considered, along with the great white, bull shark and the oceanic whitetip shark to be one of the sharks most dangerous to humans.

Taxonomy

The shark was first described by Peron and Lessueur in 1822 and was given the name *Squalus cuvier*. Muller and Henle, in 1837 renamed it *Galeocerdo tigrinus*. The genus, *Galeocerdo*, is derived from the Greek, *galeos* which means shark and the Latin *cerdus* which means the hard hairs of pigs. It is often colloquially called the leopard shark and the man-eater shark.

The tiger shark is a member of the order Carcharhiniformes; members of this order are characterised by the presence of a nictitating membrane over the eyes, two dorsal fins, an anal fin, and five gill slits. It is the largest member of the Carcharhinidae family, commonly referred to as requiem sharks. This family includes some other well known sharks such as the blue shark, lemon shark and bull shark.

Distribution

The tiger shark is often found close to the coast, in mainly tropical and subtropical waters, though they can reside in temperate waters. The shark's behaviour is primarily nomadic, but is guided by warmer currents, and it stays closer to the equator throughout the colder months. The shark tends to stay in deep waters that line reefs but does move into channels to pursue prey in shallower waters. In the western Pacific Ocean, the shark has been found as far north as Japan and as far south as New Zealand.

The shark has been recorded down to a depth of 350 metres (1,100 ft) but is also known to move into shallow water—water that would normally be considered too shallow for a species of its size. It is also frequently found in river estuaries and harbours. At night it is usually found in shallow water.

Anatomy and Appearance

Its skin can typically range from a blue or green hue to light with a white or light yellow underbelly. The distinguishing dark spots and stripes are most outstanding in young sharks and fade as the shark matures. Specimens regularly weigh 385 to 635 kg (850 to 1400 lbs). It is usually 3 metres (10 ft) to 5 metres (16 ft) long. The heaviest specimen recorded to date, a shark caught in Newcastle, NSW, Australia in 1954 and measuring a mere 5.5 metres (18 ft), scaled 1,524 kg (3,360 lb). Sexual maturity is reached at different stages for each of the sexes; males at 2.26 metres (7 ft) to 2.9 metres (10 ft) whereas females mature at 2.5 metres (8 ft) to 3.25 metres (11 ft). It has been estimated that the tiger shark can swim at a maximum speed of around 32 kilometres per hour (20 mi/h), with short bursts of higher speeds that last only a few seconds.

The tiger shark's head is somewhat wedge-shaped, which makes it easy for the shark to turn quickly to one side. Tiger sharks, as with other sharks, have small pits on the side of their upper bodies which hold electrical sensors called the ampullae of lorenzini, enabling them to detect small muscle movements of other creatures, allowing them to hunt in darkness.

In addition, the tiger shark, like many other sharks, has a mirror-like covering behind their retina called the tapidum lucidum that is exposed in darkness to reflect light that has already been seen by the retina back at it as to allow the shark to see better.

The tapidum lucidum is covered in bright light, however, as so the shark is not blinded by an excess of light. A tiger shark generally has long fins and a long upper tail; the long fins act like wings and provide lift as the shark manoeuvers through water, whereas the long tail provides bursts of speed. A tiger shark normally swims using lithe movements of its body. Its high back and dorsal fin act as a pivot, allowing it to spin quickly on its alliance.

Its teeth are flat, triangular, notched and serrated. Like most sharks, when a tiger shark loses or breaks one of its teeth, it grows a replacement tooth. The distinctive teeth seem to have evolved to be able to cut through turtle shells, and an adult tiger shark can easily bite through bone.

Diet

The tiger shark, which generally hunts at night, has a reputation for eating anything it has access to, ignoring what nutritional value the prey may or may not hold. Apart from what is thought to be sporadic feeding, its most common foods include; common fish, squid, birds, seals, other sharks, and sea turtles, alligators. The shark has a number of features which make it a good hunter, such as excellent eyesight, which allows for access to murkier waters which can offer more varieties of prey and its acute sense of smell which enable it to react to faint traces of blood in its waters and is able to follow them to the source. The tiger shark's ability to pick up on low-frequency pressure waves produced by the movements of swimming animals, for example the thrashing of an injured animal, enables the shark to find a variety of prey.

The shark is known to be aggressive. The ability to pick up low-frequency pressure waves enables the shark to advance towards an animal with confidence, even in the environment of murky water where it is often found.

The shark is known to circle its prey and even study it by prodding it with its snout. When attacking the shark devours all of its prey. Because of its aggressive nature of feeding, it is common to find a variety of foreign objects inside the digestive tract of a tiger shark. Some examples of more unusual items are automobile number plates, petroleum cans, tires, and baseballs. For this reason, the Tiger Shark is often regarded as the ocean's trash can.

Reproduction

The tiger shark breeds by internal fertilization. It is the only species in its family that is ovoviviparous; like mammals, it gives birth to live young. The male tiger shark will insert one of its claspers into the female's genital opening, acting as a guide for the sperm to be introduced. The male uses its teeth to hold the female still during the procedure, often causing the female considerable discomfort.

Mating in the northern hemisphere will generally take place between the months of March and May, with the young being born around April or June the following year. In the southern hemisphere, mating takes place in November, December, or early January.

The young are nourished inside the mothers body for up to 14 to 16 months, where the female can produce a litter ranging from 10 to 80 young. A newborn tiger shark is generally 51 centimetres (20 in) to 76 centimetres (30 in) long and leaves its mother upon birth. It is unknown how long tiger sharks live, but it has been speculated to be 20 years.

Dangers and Conservation

Although shark attacks on humans are a relatively rare phenomenon, the tiger shark is responsible for a large percentage of the fatal attacks that do occur on humans, and is regarded as one of the most dangerous species of sharks. Tiger sharks reside in temperate and tropical waters. They are often found in river estuaries and harbours, as well as shallow water close to shore, where they are bound to come into contact with humans. Because of their curious nature of feeding it is expected that a tiger shark would normally attack a human if it came in contact with it. Tiger sharks are known to dwell in waters with run-off, such as where a river enters the ocean.

Tiger sharks have become a recurring problem in Hawaii and are considered the most dangerous shark species in Hawaiian waters. They are considered to be sacred 'aumakua' or ancestor spirits by the native Hawaiians, however between 1959 and 1976, 4,668 tiger sharks were hunted down in an effort to control what was proving to be detrimental to the tourism industry. Despite these numbers, little decrease was ever detected in the attacks on humans. It is illegal to feed sharks in Hawaii and any interaction with them such as cage diving is discouraged.

While the tiger shark is not directly commercially fished, it is caught for its fins, flesh, liver, which is a valuable source of vitamin A used in the production of vitamin oils, and distinct skin, as well as by big game fishers.

Tilapia

Tilapia is the common name used for a variety of cichlid fishes of the genera *Oreochromis*, *Sarotherodon*, and *Tilapia* and is approximately equivalent to a taxonomic grouping known as the tilapiine cichlids. Tilapias inhabit a variety of fresh and, less commonly, brackish water habitats from shallow streams and ponds through to rivers, lakes, and estuaries. Most tilapias are omnivorous with a preference for soft aquatic vegetation and detritus. They have historically been of major importance in artisanal fishing in Africa and the Levant, and are of increasing importance in aquaculture around the world. Where tilapia have been deliberately or accidentally introduced, they have frequently become problematic invasive species.

Etymology

The common name tilapia is based on the name of the cichlid genus *Tilapia*, which is itself a latinisation of the Tswana word for "fish", thiape, and coined by Scottish zoologist Andrew Smith in 1840.

Other Names

Certain species of tilapia are sometimes called St. Peter's fish from the account in the Christian Bible (Matthew 17:24"27) about Peter catching a fish that carried a shekel coin in its mouth, with the dark spots on the sides of the fish being the fingermarks of the saint. Although the name is also applied to the John Dory, a marine fish not found in the Sea of Galilee, one tilapia is found there, *Sarotherodon galilaeus galilaeus*, and has been the target of small-scale artisanal fisheries for thousands of years. In some Asian countries including the Philippines, large tilapia are called pla-pla.

Tilapia As a Biological Control

Tilapias have been used as biological controls for certain aquatic plant problems. They have a preference for a floating aquatic plant, duckweed (*Lemna* sp.) but also consume some filamentous alga. These benefits are, however, frequently outweighed by the negative aspects of tilapia as invasive species. In Kenya tilapia were introduced to control mosquitoes which were causing malaria, because they consume mosquito larvae, consequently reducing the numbers of adult female mosquitos, the vector of the disease.

Tilefish

Tilefishes, also known as blanquillo, are mostly small perciform marine fish comprising the family Malacanthidae.

Commercial fisheries exist for the largest species, making them important food fish, although the American Food and Drug Administration warns pregnant women against eating them due to mercury contamination. The smaller, exceptionally colourful species are enjoyed in the aquarium.

Due to their low fecundities, commercially important species are threatened by overfishing via long-line and bottom trawling methods.

Physical Description

The two subfamilies appear to be morphologically different, with members of Branchiosteginae having deep bodies, large heads and large, somewhat subterminal mouths. In contrast, members of Malacanthinae are slender with elongate bodies, smaller heads and terminal mouths.

Tilefish range in size from 11 centimetres (yellow tilefish, *Hoplolatilus luteus*) to 125 centimetres (great northern tilefish, *Lopholatilus chamaeleonticeps*) and a weight of 30 kilograms.

Both subfamilies have long dorsal and anal fins, the latter having 1-2 spines. The gill covers (operculum) have one spine which may be sharp or blunt; some species also have a cutaneous ridge atop the head. The tail fin may range in shape from truncate to forked. Most species are fairly low-key in colour, commonly shades of yellow, brown and gray. Notable exceptions include three small, vibrant *Hoplolatilus* species: the purple sand tilefish (*H. purpureus*), Starck's tilefish (*H. starcki*) and the redback sand tilefish (*H. marcosi*).

Tilefish larvae are notable for their generous complement of spines and serrations on the head and scales. This feature also explains the family name *Malacanthidae*, from the Greek words *mala* meaning "many" and *akantha* meaning "thorn".

Habitat and Diet

Generally shallow-water fish, tilefish are usually found at depths of 50-200 metres found in both temperate and tropical waters of the Atlantic, Pacific and Indian Oceans. All species seek shelter in self-made burrows, caves at the bases of reefs or piles of rock, often in canyons or at the edges of steep slopes. Either gravelly or sandy substrate may be preferred, depending on the species.

Most species are strictly marine; an exception is found in the blue blanquillo (*Malacanthus latovittatus*) which is known to enter the brackish waters of Papua New Guinea's Goldie River.

Tilefish feed primarily on small benthic invertebrates, especially crustaceans such as crab and shrimp. Molluscs, worms, sea urchins and small fish are also taken.

Behaviour and Reproduction

Active fish, tilefish keep to themselves and generally stay at or near the bottom. They rely heavily on their keen eyesight to catch their pray. If approached, the fish will quickly dive into their constructed retreats, often headfirst. The chameleon sand tilefish (*Hoplolatilus chlupatyi*) relies on its remarkable ability to rapidly change colour (with a wide range) to evade predators.

Many species form monogamous pairs, while some are solitary in nature (e.g., ocean whitefish, *Caulolatilus princeps*), and others colonial. Some species, such as the rare pastel tilefish (*Hoplolatilus fronticinctus*) of the Indo-Pacific, actively builds large rubble mounds above which they school and in which they live. These mounds serve as both refuge and as a micro-ecosystem for other reef species.

The reproductive habits of tilefish are not well studied. Spawning occurs throughout the spring and summer; all species are presumed not to guard their broods. Eggs are small (<2 mm) and made buoyant by oil. The larvae are pelagic and drift until the fish have reached the juvenile stage.

Species

There are forty-two species in five genera.

The family is further divided into two subfamilies: Branchiosteginae or Latilinae and Malacanthinae. Some authors regard these subfamilies as two evolutionarily distinct families (in which case the former subfamily is recorded as Branchiostegidae).

- Subfamily Malacanthinae
  — Genus *Branchiostegus*

⇒ *Branchiostegus albus* (Dooley, 1978).

⇒ *Branchiostegus argentatus* (Cuvier, 1830).

⇒ *Branchiostegus auratus* (Kishinouye, 1907).

⇒ *Branchiostegus australiensis* (Dooley & Kailola, 1988).

⇒ Ribbed tilefish, *Branchiostegus doliatus* (Cuvier, 1830).

⇒ *Branchiostegus gloerfelti* (Dooley & Kailola, 1988).

⇒ *Branchiostegus hedlandensis* (Dooley & Kailola, 1988).

⇒ *Branchiostegus ilocanus* (Herre, 1928).

⇒ Red tilefish, *Branchiostegus japonicus* (Houttuyn, 1782).

⇒ *Branchiostegus paxtoni* (Dooley & Kailola, 1988).

⇒ Freckled tilefish, *Branchiostegus sawakinensis* (Amirthalingam, 1969).

⇒ Zebra tilefish, *Branchiostegus semifasciatus* (Norman, 1931).

⇒ *Branchiostegus serratus* (Dooley & Paxton, 1975).

⇒ *Branchiostegus vittatus* (Herre, 1926).

⇒ *Branchiostegus wardi* (Whitley, 1932).

— Genus *Hoplolatilus*

⇒ Chameleon sand tilefish, *Hoplolatilus chlupatyi* (Klausewitz, McCosker, Randall & Zetzsche, 1978).

⇒ Dusky tilefish, *Hoplolatilus cuniculus* (Randall & Dooley, 1974).

⇒ Yellow-spotted tilefish, *Hoplolatilus fourmanoiri* (Smith, 1964).

⇒ Pastel tilefish, *Hoplolatilus fronticinctus* (Gunther, 1887).

⇒ *Hoplolatilus geo* (Fricke & Kacher, 1982).

⇒ Yellow tilefish, *Hoplolatilus luteus* (Allen & Kuiter, 1989).

⇒ Redback sand tilefish, *Hoplolatilus marcosi* (Burgess, 1978).

⇒ *Hoplolatilus oreni* (Clark & Ben-Tuvia, 1973).

⇒ *Hoplolatilus pohle* (Earle & Pyle, 1997).

⇒ Purple sand tilefish, *Hoplolatilus purpureus* (Burgess, 1978).

⇒ Starck's tilefish, *Hoplolatilus starcki* (Randall & Dooley, 1974).

— Genus *Malacanthus*

⇒ Quakerfish, *Malacanthus brevirostris* (Guichenot, 1848).

⇒ Blue blanquillo, *Malacanthus latovittatus* (Lacepede, 1801).

⇒ Sand tilefish, *Malacanthus plumieri* (Bloch, 1786).

• Subfamily Latilinae

— Genus *Caulolatilus*

⇒ Bighead tilefish, *Caulolatilus affinis* (Gill, 1865).

⇒ Bermuda tilefish, *Caulolatilus bermudensis* (Dooley, 1981).

⇒ Atlantic goldeye tilefish, *Caulolatilus chrysops* (Valenciennes, 1833).

⇒ Blackline tilefish, *Caulolatilus cyanops* (Poey, 1866).

⇒ Bankslope tilefish, *Caulolatilus dooleyi* (Berry, 1978).

⇒ Reticulated tilefish, *Caulolatilus guppyi* (Beebe & Tee-Van, 1937).

⇒ Hubbs' tilefish, *Caulolatilus hubbsi* (Dooley, 1978).

⇒ Gulf bareye tilefish, *Caulolatilus intermedius* (Howell Rivero, 1936).

⇒ Grey tilefish, *Caulolatilus microps* (Goode & Bean, 1878).

⇒ Ocean whitefish, *Caulolatilus princeps* (Jenyns, 1840).

⇒ Yellowbar tilefish, *Caulolatilus williamsi* (Dooley & Berry, 1977).

— Genus *Lopholatilus*

⇒ Great northern tilefish, *Lopholatilus chamaeleonticeps* (Goode & Bean, 1879) (known as Golden Tile culinarily)

⇒ *Lopholatilus villarii* (Miranda-Ribeiro, 1915).

Toadfish

The toadfish are family Batrachoididae is the only family in the of ray-finned fish order Batrachoidiformes. They are so-named because of their rather drab colouration reminiscent of that of terrestrial toads (*batrachus* is Greek for frog). There are 69 species in 19 genera, most of which are marine in distribution though some are found in brackish water and one subfamily, the Thalassophryninae, is found exclusively in freshwater habitats in South America.

Toadfish are benthic ambush predators that favour sandy or muddy substrates where their cryptic colouration helps them avoid detection by their prey.

The dorsal fin and gill cover spines on the toadfishes of the subfamily Thalassophryninae are hollow and will inject venom into any predator attempting to eat the fish.

Toadfish are well known for their ability to "sing", males in particular using the swim bladder as a sound-production device used to attract mates. The Western Atlantic species *Opsanus tau* known as the oyster toadfish is quite widely used as a research animal, while a few species, most notably *Thalassophryne amazonica*, are occasionally kept as aquarium fish.

Morphology

Toadfishes are usually scaleless, with eyes set high on large heads. Their mouths are also large, with both maxilla and premaxilla. The gills are small and occur only on the sides of the fish. The pelvic fins are forward of the pectoral fins, usually under the gills, and have one spine with several soft rays. Three are two separate dorsal fins, the first smaller dorsal fin with spines; and the second larger and longer dorsal, with from 15 to 25 soft rays. The number of vertebra range from 25 to 47.

Toadfishes of the genus *Porichthys*, the midshipman fishes, have photophores and four lateral lines, while the Thalassophryninae are venomous, with a total of four hollow spines (two dorsal and one on each gill-flap (opercle)) connecting to venom glands and capable of delivering a painful wound.

Distribution

Toadfishes are found worldwide. Almost all are marine, but *Daector quadrizonatus* and *Thalassophryne amazonica* are known from Colombia (Atrato River) and the Amazon River, respectively.

Habits

Toadfishes are bottom-dwellers, ranging from near shore areas to deep waters. They tend to be omnivorous, eating sea worms, crustaceans, molluscs and other fish. They often hide in rock crevices, among the bottom vegetation, or even dig dens in the bottom sediments, from which they ambush their prey.

Males make the nests and guard them after the female lays the eggs. The male attracts the female by "singing", that is by releasing air by contracting muscles on their swim bladder. The sound has been called a 'hum' or 'whistle'.

Economics

Toadfish are not normally commercially exploited, however, they are taken by local fishermen as a food fish, and by trawlers where they usually end up as a source of fishmeal and oil. Some smaller toadfish from brackish-water habitats have been exported as freshwater aquarium fishes.

Tonguefish

Tonguefishes are a family, Cynoglossidae, of flatfishes.

They are found in tropical and subtropical oceans, mainly in shallow waters and estuaries, though a few species found in deep sea floors, and a few in rivers.

Some species have been observed congregating around ponds of sulphur that pool up from beneath the seafloor. Scientists are unsure of the mechanism that allows the fish to survive and even thrive in such a hostile environment.

Species

There are 137 species in 3 genera:

- Genus *Cynoglossus*
  - — Three-lined tongue sole, *Cynoglossus abbreviatus* (Gray, 1834).
  - — Natal tongue-fish, *Cynoglossus acaudatus* (Gilchrist, 1906).
  - — Sharpnose tonguesole, *Cynoglossus acutirostris* (Norman, 1939).
  - — Largescale tonguesole, *Cynoglossus arel* (Bloch & Schneider, 1801).
  - — Fourline tonguesole, *Cynoglossus attenuatus* (Gilchrist & Thompson, 1917).
  - — Fourlined tonguesole, *Cynoglossus bilineatus* (Lacepede, 1802).
  - — Southern tongue sole, *Cynoglossus broadhursti* (Waite, 1905).
  - — Nigerian tonguesole, *Cynoglossus browni* (Chabanaud, 1949).
  - — Ghanian tonguesole, *Cynoglossus cadenati* (Chabanaud, 1947).
  - — Canary tonguesole, *Cynoglossus canariensis* (Steindachner, 1882).
  - — Sand tonguefish, *Cynoglossus capensis* (Kaup, 1858).
  - — Hooked tonguesole, *Cynoglossus carpenteri* (Alcock, 1889).
  - — Bengal tongue sole, *Cynoglossus cynoglossus* (Hamilton, 1822).
  - — Roundhead toungesole, *Cynoglossus dispar* (Day, 1877).
  - — *Cynoglossus dollfusi* (Chabanaud, 1931).

— Carrot tonguesole, *Cynoglossus dubius* (Day, 1873).

— Durban tonguesole, *Cynoglossus durbanensis* (Regan, 1921).

— River tonguesole, *Cynoglossus feldmanni* (Bleeker, 1853).

— Ripplefin tonguesole, *Cynoglossus gilchristi* (Regan, 1920).

— *Cynoglossus gracilis* (Gunther, 1873).

— Freshwater tongue sole, *Cynoglossus heterolepis* (Weber, 1910).

— Genko sole, *Cynoglossus interruptus* (Gunther, 1880).

— *Cynoglossus itinus* (Snyder, 1909).

— Red tonguesole, *Cynoglossus joyneri* (Gunther, 1878).

— *Cynoglossus kapuasensis* (Fowler, 1905).

— Shortheaded tonguesole, *Cynoglossus kopsii* (Bleeker, 1851).

— Lachner's tonguesole, *Cynoglossus lachneri* (Menon, 1977).

— Roughscale tonguesole, *Cynoglossus lida* (Bleeker, 1851).

— *Cynoglossus lighti* (Norman, 1925).

— *Cynoglossus lineolatus* (Steindachner, 1867).

— Long tongue sole, *Cynoglossus lingua* (Hamilton, 1822).

— *Cynoglossus maccullochi* (Norman, 1926).

— *Cynoglossus macrolepidotus* (Bleeker, 1851).

— Big-eyed tongue-sole, *Cynoglossus macrophthalmus* (Norman, 1926).

— Malabar tonguesole, *Cynoglossus macrostomus* (Norman, 1928).

— *Cynoglossus maculipinnis* (Rendahl, 1921).

— Threeline tonguesole, *Cynoglossus marleyi* (Regan, 1921).

— *Cynoglossus melampetalus* (Richardson, 1846).

— Smallscale tonguesole, *Cynoglossus microlepis* (Bleeker, 1851).

— Guinean tonguesole, *Cynoglossus monodi* (Chabanaud, 1949).

— *Cynoglossus monopus* (Bleeker, 1849).

— *Cynoglossus nigropinnatus* (Ochiai, 1963).

— *Cynoglossus ogilbyi* (Norman, 1926).

— *Cynoglossus oligolepis* (Bleeker, 1854).

— *Cynoglossus pottii* (Steindachner, 1902).

— Speckled tonguesole, *Cynoglossus puncticeps* (Richardson, 1846).

— *Cynoglossus purpureomaculatus* (Regan, 1905).

— *Cynoglossus robustus* (Gunther, 1873).

— *Cynoglossus roulei* (Wu, 1932).

— *Cynoglossus sealarki* (Regan, 1908).

— Bengal tongue-sole, *Cynoglossus semifasciatus* (Day, 1877).

— Tongue sole, *Cynoglossus semilaevis* (Gunther, 1873).

— Senegalese tonguesole, *Cynoglossus senegalensis* (Kaup, 1858).

— *Cynoglossus sibogae* (Weber, 1913).

— *Cynoglossus sinicus* (Wu, 1932).

— *Cynoglossus sinusarabici* (Chabanaud, 1931).

— *Cynoglossus suyeni* (Fowler, 1934).

— *Cynoglossus trigrammus* (Gunther, 1862).

— Macau sole, *Cynoglossus trulla* (Cantor, 1849).

— *Cynoglossus waandersii* (Bleeker, 1854).

— Zanzibar tonguesole, *Cynoglossus zanzibarensis* (Norman, 1939).

• Genus *Paraplagusia*

— Doublelined tonguesole, *Paraplagusia bilineata* (Lacepede, 1802).

— Bloch's tonguesole, *Paraplagusia blochii* (Bleeker, 1851).

— *Paraplagusia guttata* (Macleay, 1878).

— Black cow-tongue, *Paraplagusia japonica* (Temminck & Schlegel, 1846).

— Long-snouted tongue sole, *Paraplagusia longirostris* (Chapleau, Renaud & Kailola, 1991).

— Dusky tongue sole, *Paraplagusia sinerama* (Chapleau & Renaud, 1993).

- Genus *Symphurus*

— Caribbean tonguefish, *Symphurus arawak* (Robins & Randall, 1965).

— Inkspot tonguefish, *Symphurus atramentatus* (Jordan & Bollman, 1890).

— California tonguefish, *Symphurus atricaudus* (Jordan & Gilbert, 1880).

— *Symphurus australis* (Rendahl, 1922).

— *Symphurus bathyspilus* (Krabbenhoft & Munroe, 2003).

— Kriete's tonguefish, *Symphurus billykrietei* (Munroe, 1998).

— Chocolate tonguefish, *Symphurus callopterus* (Munroe & Mahadeva, 1989).

— *Symphurus caribbeanus* (Munroe, 1991).

— Chabanaud's tonguefish, *Symphurus chabanaudi* (Mahadeva & Munroe, 1990).

— Offshore tonguefish, *Symphurus civitatium* (Ginsburg, 1951).

— Devil's tonguefish, *Symphurus diabolicus* (Mahadeva & Munroe, 1990).

— Spottedfin tonguefish, *Symphurus diomedeanus* (Goode & Bean, 1885).

— Elongate tonguefish, *Symphurus elongatus* (nil).

— Banded tongue-fish, *Symphurus fasciolaris* (Gilbert, 1892).

— *Symphurus fuscus* (Brauer, 1906).

— *Symphurus gilesii* (Alcock, 1889).

— Ginsburg's tonguefish, *Symphurus ginsburgi* (Menezes & Benvegnu, 1976).

— Gorgonian tonguefish, *Symphurus gorgonae* (Chabanaud, 1948).

— *Symphurus hondoensis* (Hubbs, 1915).

— *Symphurus insularis* (Munroe, Brito & Hernandez, 2000).

— Jenyn's tonguefish, *Symphurus jenynsi* (Evermann & Kendall, 1907).

— *Symphurus kyaropterygium* (Menezes & Benvegnu, 1976).

— Lee's tonguefish, *Symphurus leei* (Jordan & Bollman, 1890).

— Elongate tonguesole, *Symphurus ligulatus* (Cocco, 1844).

— *Symphurus lubbocki* (Munroe, 1990).

— *Symphurus luzonensis* (Chabanaud, 1955).

— *Symphurus macrophthalmus* (Norman, 1926).

— *Symphurus maldivensis* (Chabanaud, 1955).

— Margined tonguefish, *Symphurus marginatus* (Goode & Bean, 1886).

— *Symphurus marmoratus* (Bleeker, 1851).

— Drab tonguefish, *Symphurus melanurus* (Clark, 1936).

— Blackstripe tonguefish, *Symphurus melasmatotheca* (Munroe & Nizinski, 1990).

— Smallfin tonguefish, *Symphurus microlepis* (Bleeker, 1851).

— *Symphurus microrhynchus* (Weber, 1913).

— Largescale tonguefish, *Symphurus minor* (Ginsburg, 1951).

— *Symphurus monostigmus* (Munroe, 2006).

— Freckled tonguefish, *Symphurus nebulosus* (Goode & Bean, 1883).

— Tonguesole, *Symphurus nigrescens* (Rafinesque, 1810).

— Norman's tonguesole, *Symphurus normani* (Chabanaud, 1950).

— *Symphurus novemfasciatus* (Shen & Lin, 1984).

— Double-spot tonguesole, *Symphurus ocellatus* (nil).

— *Symphurus oculellus* (Munroe, 1991).

— Spotfin tonguefish, *Symphurus oligomerus* (Mahadeva & Munroe, 1990).

— Ocellated tonguefish, *Symphurus ommaspilus* (Bohlke, 1961).

— *Symphurus orientalis* (Bleeker, 1879).

— Pygmy tonguefish, *Symphurus parvus* (Ginsburg, 1951).

— Longtail tonguefish, *Symphurus pelicanus* (Ginsburg, 1951).

— Deepwater tonguefish, *Symphurus piger* (nil).

— (Linnaeus, 1766).

— Duskycheek tonguefish, *Symphurus plagusia* (Bloch & Schneider, 1801).

— Halfstriped tonguefish, *Symphurus prolatinaris* (Munroe, Nizinski & Mahadeva, 1991).

— Northern tonguefish, *Symphurus pusillus* (Goode & Bean, 1885).

— *Symphurus regani* (Weber & de Beaufort, 1929).

— *Symphurus reticulatus* (Munroe, 1990).

— Patchtail tonguefish, *Symphurus rhytisma* (Bohlke, 1961).

— *Symphurus schultzi* (Chabanaud, 1955).

— Sevenband tonguesole, *Symphurus septemstriatus* (Alcock, 1891).

— Blotchfin tonguefish, *Symphurus stigmosus* (Munroe, 1998).

— Blackbelly tonguesole, *Symphurus strictus* (Gilbert, 1905).

— *Symphurus tessellatus* (Quoy & Gaimard, 1824).

— Trewavas' tonguefish, *Symphurus trewavasae* (Chabanaud, 1948).

— Threeband tonguesole, *Symphurus trifasciatus* (Alcock, 1894).

— *Symphurus undatus* (Gilbert, 1905).

— Dark cheek tongue, *Symphurus undecimplerus* (Munroe & Nizinski, 1990).

— Spottail tonguefish, *Symphurus urospilus* (Ginsburg, 1951).

— *Symphurus vanmelleae* (Chabanaud, 1952).

— *Symphurus variegatus* (Gilchrist, 1903).

— Mottled tonguefish, *Symphurus varius* (Garman, 1899).

— William's tonguefish, *Symphurus williamsi* (Jordan & Culver, 1895).

— *Symphurus woodmasoni* (Alcock, 1889).

Torrent Fish

The Torrent Fish, *Cheimarrichthys fosteri*, is the only member of the genus Cheimarrichthys which in turn is the only member of the family Cheimarrichthyidae. It is found only in New Zealand, in stony rivers and streams all around the North and South Islands. It grows to a maximum length of 15 cm.

The torrentfish lives in the swift whitewater rapids of stony rivers and streams, and adaptations to enable it to live in this environment include a flattened head, large pectoral fins which help it to anchor on the riverbed, raised eyes, and a ventral mouth. Despite its skill at living in swift water however, the torrentfish is not a good climber and only penetrates inland in river systems where the gradient is relatively low.

Like many of New Zealand's freshwater fish, the torrentfish is amphidromous and undertakes migrations between the sea and fresh water as part of its life cycle. Juvenile torrentfish enter fresh water in spring and autumn, and after a few weeks in the estuaries, begin moving upstream to the adult habitat. The adults continue to move slowly upstream, with the largest and most inland fish being the females and those in the lower reaches predominantly males. The spawning behaviour is unknown.

Treefish

The treefish (*Sebastes serriceps*) is a marine fish of the *Sebastes* genus. It is native to the eastern Pacific Ocean with a range from San Francisco, California to central Baja California, Mexico. It has been reported up to 41 cm, and usually occurs in crevices in rocky areas.

Triggerfish

Triggerfishes are brightly coloured fish of the family Balistidae. Often marked by lines and spots, they inhabit warm coastal waters of the Atlantic, Mediterranean and the Indo-Pacific.

Anatomy and Appearance

Their size varies from 30 cm (1 foot) to 75 cm (2.5 feet). Triggerfish have a roundish, laterally flat body with an anterior dorsal fin. They can erect the first two dorsal spines, the first one locks and the second one unlocks. This prevents predators from swallowing them or pulling them out of their holes. This locking and unlocking behaviour is why they are named 'triggerfish'.

They have a small pectoral fin, fused to one spine. Unlike the spine of a filefish, the spine of the triggerfish can be held in place by a second spine to make the fish more threatening to the predator. Their small eyes, situated on top of their large head, can be rotated independently.

They have tough skin, covered with rough rhomboid-shaped scales that form a tough armour on their body. A big, angular-shaped head extends into a snout with strong jaws and sharp teeth, made for crushing shells. Each jaw contains a row of eight teeth, while the upper jaw contains another set of six plate-like teeth.

Behaviour

Most are solitary and diurnal. They feed on hard-shelled invertebrates, a few feed on large zooplankton or algae. They lay their demersal eggs in a small hole, dug in the ground. Some species guard their eggs.

A few of the triggerfish species can be quite aggressive during reproduction season. In particular Picasso triggerfish and titan triggerfish viciously defend their circular nests against any intruders, including scuba divers and snorkellers. Their territory extends in a cone shape from the nest to the surface, so swimming upwards puts one further into the fishes' territory. A horizontal swim away from the nest site is the most sensible course of action when confronted by an angry triggerfish. In contrast to the relatively small Picasso triggerfish, the titan triggerfish poses a serious threat to inattentative divers due to its large size and powerful teeth.

Some species of triggerfish are known to make a sound akin to a grunt or snarl when taken out of the water.

Genera.

- *Abalistes* (Bloch & Schneider, 1801)
  - — *Abalistes stellaris* : Starry triggerfish
- *Balistapus* (Tilesius, 1820)
  - — *Balistapus undulatus* : Orange-lined triggerfish
- *Balistes* (Linnaeus, 1758)
  - — *Balistes capriscus*: Grey triggerfish
  - — *Balistes ellioti*
  - — *Balistes polylepis* : Finescale triggerfish
  - — *Balistes punctatus* : Bluespotted triggerfish
  - — *Balistes rotundatus*
  - — *Balistes vetula* : Queen triggerfish
  - — *Balistes willughbeii*
- *Balistoides* (Fraser-Brunner, 1935)
  - — *Balistoides conspicillum* : Clown triggerfish

— *Balistoides viridescens* : Titan triggerfish or mustache triggerfish

- *Canthidermis* (Swainson, 1839)

— *Canthidermis maculata*: Spotted oceanic triggerfish

— *Canthidermis macrolepis* : Large-scale triggerfish

— *Canthidermis sufflamen* : Ocean triggerfish

- *Melichthys* (Swainson, 1839)

— *Melichthys indicus* : Indian triggerfish

— *Melichthys niger* : Black triggerfish

— *Melichthys vidua* : Pinktail triggerfish

- *Odonus* (Gistel, 1848)

— *Odonus niger* : Redtoothed triggerfish

- *Pseudobalistes* (Bleeker, 1865)

— *Pseudobalistes flavomarginatus:* Yellowmargin triggerfish

— *Pseudobalistes fuscus* : Blue or rippled triggerfish, blue-and-gold triggerfish or yellowspotted triggerfish

— *Pseudobalistes naufragium* : Stone triggerfish

- *Rhinecanthus* (Swainson, 1839)

— *Rhinecanthus abyssus*

— *Rhinecanthus aculeatus:* Picasso triggerfish, lagoon triggerfish *(USA)*

— *Rhinecanthus assasi* : Arabian picasso triggerfish

— *Rhinecanthus cinereus*

— *Rhinecanthus lunula* : Halfmoon picassofish

— *Rhinecanthus rectangulus*: Reef triggerfish or humuhumu-nukunuku-a-puaa *(Hawaii)*

— *Rhinecanthus verrucosus* : Blackbelly triggerfish

- *Sufflamen* (Jordan, 1916)

— *Sufflamen albicaudatum* : Bluethroat triggerfish

— *Sufflamen bursa* : Boomerang triggerfish

— *Sufflamen chrysopterum* : Halfmoon triggerfish

— *Sufflamen fraenatum* : Masked triggerfish
— *Sufflamen verres* : Orangeside triggerfish

- *Xanthichthys* (Kaup in Richardson, 1856)
  — *Xanthichthys auromarginatus* : Gilded triggerfish
  — *Xanthichthys caeruleolineatus* : Bluelined triggerfish
  — *Xanthichthys lineopunctatus* : Striped triggerfish
  — *Xanthichthys mento* : Redtail triggerfish
  — *Xanthichthys ringens* : Sargassum triggerfish
- *Xenobalistes* (Matsuura, 1981)
  — *Xenobalistes punctatus* : Outrigger triggerfish
  — *Xenobalistes tumidipectoris*

Threefin Blenny

Threefin or triplefin blennies are blennioids, small perciform marine fish of the family Tripterygiidae. Found in tropical and temperate waters of the Atlantic, Pacific and Indian Oceans, the family contains approximately 150 species in 30 genera. The family name derives from the Greek *tripteros* meaning "with three wings".

With an elongate, typical blenny form, threefin blennies differ from their relatives by having a dorsal fin separated into three parts (hence the name); the first two are spinous. The small, slender pelvic fins are located underneath the throat and possess a single spine; the large anal fin may have one or two spines. The pectoral fins are greatly enlarged, and the tail fin is rounded. The New Zealand topknot, *Notoclinus fenestratus*, is the largest species at 20 cm in total length; most other species do not exceed 6 cm.

Many threefin blennies are brightly coloured, often for reasons of camouflage; these species are popular in the aquarium hobby. As benthic fish, threefin blennies spend most of their time on or near the bottom on coral and rocks. The fish are typically found in shallow, clear waters with sun exposure, such as lagoons and seaward reefs; nervous fish, they retreat to rock crevices at any perceived threat.

Threefin blennies are diurnal and territorial; many species exhibit sexual dichromatism, with the females drab compared to the males. The second dorsal fin is also extended in the males of some species. Small invertebrates comprise the bulk of the threefin blenny diet.

Genera

- *Acanthanectes*
- *Apopterygion*
- *Axoclinus*
- *Bellapiscis*
- *Blennodon*
- *Brachynectes*
- *Ceratobregma*

Triacanthidae

Triplespines (Tracanthidae) are a family in the order of the Tetraodontiformes, which consists of seven species in seven genera, in addition to one extinct genus. Much like their relatives the triggerfish and the filefish, the triplespines's first ray of the dorsal fin is formed to a spine.

Further, they have two spines in place of their ventral fins. Not much is known about how the fish live. They live in the Indo-Pacific Ocean. All but the *Triacanthus biaculeatus* are off shore fish, that only come close to land occasionally. *Triacanthus biaculeatus* lives from the Persian Gulf along the Asian coast all the way to Japan, as well as off the coast of Australia. They can get up to 15 to 25 centimetres long.

Tripod Fish

The tripod fish, *Bathypterois grallator*, is an unusual bathypelagic (deep sea) fish and is named for the long extensions of its pelvic and lower caudal fins, on which it stands on the sea floor. The tripod fish is closely related to the smaller spiderfish (*Bathypterois longifilis*), which is similar in appearance and habits but smaller and with much shorter fin extensions;

the two species are often found standing very near to one another on the ocean floor.

Description

Tripod fish are relatively small, the largest known specimen having measured only 37 cm (14.5 in); however the three elongated fins of the tripod fish may extend to nearly one metre (3 ft 3 in) in length. The fish is slender, deeper than it is wide and with very small eyes that probably are not useful at the depths at which the fish lives. Tripod fish are very sensitive to the vibrations of other animals in the water. In addition to its tripods, the fish also has unusually large pectoral fins. Tripod fish have been found at depths of anywhere from 900m to 3500m (2950-11500 ft), and are distributed in all oceans in the equatorial regions.

The tripod fish is a relatively sedentary fish. It spends much of its adult life standing on the ocean bottom on its fins. The fish stands facing the prevailing current, and hunts by extending its unusually long pectoral fins into the current and waiting for the small crustaceans on which it feeds to simply bump into its fins. The fish grasps its prey in the pectoral fins and directs it towards its mouth. The extensions of the pelvic and caudal fins are stiff enough for the fish to stand on them for (presumably) extended periods of time. However, deep sea researchers have succeeded in surprising the fish enough to make it swim; when it swims, the tripods seem to be quite flexible.

Hermaphroditism

All tripod fish exhibit both male and female sex organs. While this trait (hermaphroditism) is not uncommon in other animals of the deep sea, the tripod fish is unusual in that the male and female organs reach maturity at the same time, thus allowing the tripod fish to fertilize its own eggs. It is supposed that tripod fish may be so sparsely distributed through the oceans that one fish may not be able to find another to mate when the time comes; thus as a last resort a single tripod fish could still reproduce.

Life Cycle

Very little is known about the life cycle of the tripod fish. The habits of the young of the species are virtually unknown. It is estimated that the species has very low resilience, and based on its sexual habits the total population and genetic variance of the tripod fish are both assumed to be small.

Trout

Trout is the common name given to a number of species of freshwater fish belonging to the salmon family, *Salmonidae*.

All fish called trout are members of the subfamily *Salmoninae*. The name is commonly used for species in three of the seven genera in the subfamily: *Salmo*, which includes Atlantic species; *Oncorhynchus*, which includes Pacific species; and *Salvelinus*, which includes fish also sometimes called *char* or *charr*. Fish referred to as trout include:

- Genus *Salmo*
  - — Adriatic trout, *Salmo obtusirostris*
  - — Brown trout, *Salmo trutta*
  - — Flathead trout, *Salmo platycephalus*
  - — Marmorata, Soca River trout or Soea trout - *Salmo trutta marmoratus*
  - — Ohrid trout, *Salmo letnica*
  - — Sevan trout, *Salmo ischchan*
- Genus *Oncorhynchus*
  - — Apache trout, *Oncorhynchus apache*
  - — Seema, *Oncorhynchus masou*
  - — Cutthroat trout, *Oncorhynchus clarki*
  - — Gila trout, *Oncorhynchus gilae*
  - — Golden trout, *Oncorhynchus aguabonita*
  - — Rainbow trout, *Oncorhynchus mykiss*
  - — Mexican Golden Trout, *Oncorhynchus chrysogaster* and as many as eight other species or sub-species in northwest Mexico, not yet formally named.

- Genus *Salvelinus* (Char)
  - — Arctic char, *Salvelinus alpinus*
  - — Aurora trout, *Salvelinus fontinalis timagamiensis*
  - — Brook trout, *Salvelinus fontinalis*
  - — Bull trout, *Salvelinus confluentus*
  - — Dolly Varden trout, *Salvelinus malma*
  - — Lake trout, *Salvelinus namaycush*
  - — Silver trout, *Salvelinus fontinalis agassizi* (extinct)

Trout are usually found in cool, clear streams and lakes, although many of the species have anadromous strains as well. They are distributed naturally throughout North America, northern Asia and Europe. Several species of trout were introduced to Australia and New Zealand by amateur fishing enthusiasts in the 19th century, effectively displacing and endangering several upland native fish species. The introduced species included brown trout from England, and rainbow trout from California. The rainbow trout were a steelhead strain, generally accepted as coming from Sonoma Creek. The rainbow trout of New Zealand still show that steelhead tendency to run up rivers in winter to spawn.

Trout have fins entirely without spines, and all of them have a small adipose (fatty) fin along the back, near the tail. There are many species, and even more populations that are isolated from each other and morphologically different. However, many of these distinct populations show no significant genetic differences, and therefore what may appear to be a large number of species is considered a much smaller number of distinct species by most ichthyologists.

The trout found in the eastern United States are a good example of this. The brook trout, the aurora trout and the (extinct) silver trout all have physical characteristics and colourations that distinguish them, yet genetic analysis shows that they are one species, *Salvelinus fontinalis*.

Lake trout (*Salvelinus namaycush*), like brook trout, actually belong to the char genus. Lake trout inhabit many of the larger lakes in North America, and live much longer than rainbow trout, which have an average maximum lifespan of 7 years. Lake trout can live many decades, and can grow to more than 30 kg (66 pounds).

Trout generally feed on soft bodied aquatic invertebrates, such as flies, mayflies, caddisflies, stoneflies, and dragonflies. In lakes, various species of zooplankton often form a large part of the diet. In general, trout longer than about 30 cm prey almost exclusively on fish, where they are available. Adult trout will devour fish exceeding 1/3 their length.

As a group, trout are somewhat bony, but the flesh is generally considered to be tasty. Additionally, they provide a good fight when caught with a hook and line, and are sought after recreationally. Because of their popularity, trout are often raised on fish farms and planted into heavily fished waters, in an effort to mask the effects of overfishing. While they can be caught with a normal rod and reel, fly fishing is a distinctive method developed primarily for trout, and now extended to other species. Farmed trout and char are also sold commercially as food fish.

Trout that live in different environments can have dramatically different colourations and patterns. Mostly, these colours and patterns form as camouflage, based on the surroundings, and will change as the fish moves to different habitats. Trout in, or newly returned from the sea, can look very silvery, while the same "genetic" fish living in a small stream or in an alpine lake could have pronounced markings and more vivid colouration. It is virtually impossible to define a particular colour pattern as belonging to a specific breed; however, in general, wild fish are claimed to have more vivid colours and patterns.

The cutthroat trout has 14 recognised subspecies (depending on your sources), such as the Lahontan cutthroat trout, *Oncorhynchus clarki henshawi*, Bonneville cutthroat trout, *Oncorhynchus clarki utah*, Colorado River cutthroat trout, Yellowstone cutthroat trout.

Trout Cod

The trout cod, *Maccullochella macquariensis,* is a large and striking predatory freshwater fish of the *Maccullochella* genus and the Percichthyidae family which was originally found in the southeast corner of the Murray-Darling river system in Australia. It is closely related to the Murray cod.

In the 1800s and early 1900s, when trout cod were widely recognised as a separate species, they were generally known as blue-nose cod or simply blue-nose, particularly in Victoria. In some parts of New South Wales however they were also known as trout cod, and this common name was adopted when the species status of the fish was finally confirmed by genetic studies in the early 1970s. This choice of official common name was perhaps unfortunate; it has been suggested that blue-nose cod is a more appropriate name as the name trout cod causes confusion amongst the Australian public.

Trout cod are a listed species on a number of different registers including Endangered under the New South Wales Fisheries Management Act 1994, the Australian Commonwealth's Environment Protection and Biodiversity Conservation Act 1999, the Australian Capital Territory's Nature Conservation Act 1980 and by the World Conservation Union (IUCN). They are also listed as Threatened under the Victorian Flora and Fauna Guarantee Act 1998.

Description

Trout cod have been reliably recorded to at least 80 cm and 16 kg, but there are some credible anecdotal accounts of larger specimens.

Trout cod are broadly similar to the Murray cod, however there are some distinct differences in morphology and colouration.

Trout cod are a small to medium groper-like fish with a deep, elongated body that is round in cross-section. In contrast to Murray cod, trout cod have a pointed head with the top jaw overhanging the bottom jaw, and the slope of the head is straight. The eyes are slightly larger and more prominent than in Murray cod. The head tends to be free of speckling however a distinct dark stripe through the eye is usually present.

Trout cod are cream to light grey on their ventral ("belly") surfaces. Their back and flanks are most commonly bluish-grey in colour, overlain with irregular black speckling, but this can be highly variable depending on the habitat specimens come from, and can range from almost white to light grey-green, light brown, dark brown or almost black. The black speckling on the back and flanks is consistent however.

The spiny dorsal fin is moderate in height and is partially separated by a notch from the high, rounded soft dorsal fin. Soft dorsal, anal and caudal (tail) fins are all large and rounded, and are light grey to dark grey or black with distinct white edges. The large, rounded pectoral fins are usually similar in colour to flanks. The pelvic fins are large and angular and set forward of the pectoral fins. The leading white-coloured rays on the pelvic fins split into two trailing white filaments, while the pelvic fins themselves are usually a translucent cream or light grey.

Smaller trout cod tend to be more slender than equivalent sized Murray cod; curiously, very large trout cod appear to develop deeper shoulders than equivalent sized Murray cod.

While trout cod were only conclusively described as a separate species to Murray cod in 1972, commercial and recreational fishermen and early fishery biologists were in no doubt that there were two separate species of cod in the Murray-Darling system from the 1850s onwards, and noted the trout cod's different appearance and spawning biology and preference for cooler, faster flowing water and upland habitats. During the 1800s and early 1900s trout cod were recognised by the scientific community as separate species, due to differing habitat preferences, morphological differences (especially the much smaller size at sexual maturity) and differing spawning times. It was really only post- World War II — by which time trout cod had become very rare or even extinct in much of their original range — that the erroneous idea that they were really just anomalous Murray cod gained any currency.

Habitat

Although there is/was a very substantial overlap in range, trout cod are essentially a more specialised upland sister species to Murray cod. Therefore, the trout cod's main habitats were the larger upland rivers and creeks, which they usually co-inhabited with Macquarie perch and one or both of the blackfish species. Historical research is confirming a primarily upland distribution for trout cod; recent governmental literature lacking such historical research and suggesting trout cod are primarily a lowland fish species must be considered inaccurate.

Division into specialist upland and *primarily* lowland species is a relatively common phenomena in native fish genera of the Murray-Darling and East Coast systems with other notable pairs are shown in the following table:

Upland	*Lowland*
Macquarie perch, *Macquaria australasica*	Golden perch, *Macquaria ambigua*
Australian bass, *Macquaria novemaculeata*	Estuary perch, *Macquaria colonorum*
Two-spined blackfish, *Gadopsis bispinosis*	River blackfish, *Gadopsis marmoratus*
Mountain galaxias species complex, *Galaxias spp.*	Flathead galaxias, *Galaxias rostratus*

Trout cod are often found close to cover and in faster currents and in cooler waters than Murray cod. Their diet is essentially the same as Murray cod with adjustment made for size, eating mainly other fishes, freshwater mussels, crustaceans, aquatic insects, small mammals and water fowl. However, recent anecdotal evidence suggests terrestrial insects made up a significant proportion of the trout cod's diet in upland rivers and streams.

In the surviving Murray River population trout cod tend to stick to areas of deep water near banks, around snags, rocks or other large structure. However, historical accounts of trout cod in upland river habitats stated that trout cod were often found in shallow riffles and runs. Generally speaking, radio-tracked trout cod in the surviving Murray River population have small home ranges and may be a species which does not move away from their original base, except during the breeding season when they follow a common trend in Murray-Darling fish of migrating upstream prior to spawning. It seems likely that trout cod follow a similar pattern to Murray cod and return post- spawning to their original location.

Diet

Trout cod are carnivores and feed on other fish, crustaceans (such as crayfish, yabbies and freshwater shrimp) as well as aquatic and terrestrial insects.

Reproduction

Trout cod reach sexual maturity at 3 to 5 years (which corresponds to about 35 cm in males and 43 cm in females). Trout cod reach sexual maturity at a smaller size than Murray cod, which is an adaptation to the rocky, low nutrient and often quite small upland habitats trout cod were found in. Spawing of trout cod has never been observed in the wild and is not well understood. It is believed to be essentially the same as Murray cod but occurs about three weeks earlier and at significantly lower temperatures in waters shared by the two species. Trout cod are believed to spawn at temperatures as low as 15 degrees in upland rivers, using rocks as a spawning substrate; these are also clear adaptations to cool, rocky upland river habitats. Significantly, and unlike Murray cod, trout cod will not breed in earthen dam brood ponds; another indication that trout cod are a more specialised upland species than Murray cod. Artificial breeding programmes being conducted for the species recovery use hormone injections to induce ovulation in naturally ripe fish in spring.

Conservation

Trout cod were once quite abundant, but are now gravely threatened by overfishing, environmental factors and introduced trout species and are listed as endangered. The species is now totally protected. Only one wild, naturally occurring trout cod population remains in the Murray River in a region where the river is basically an extended transition zone from upland river habitat to lowland river habitat.

In more lowland river habitats, river regulation and habitat degradation through activities like desnagging, and overfishing, are probably the primary causes of decline. The possibility that trout cod populations in upland habitats were the ultimate source of trout cod populations in lowland habitats (i.e. source and sink populations) cannot be discounted however.

Historical accounts such as those from J.O. Langtry indicate lowland trout cod populations were secondary populations in secondary habitats, clearly in the minority to more abundant primarily lowland native fish species such as Murray cod, golden perch and silver perch. Therefore it is doubtful whether strong trout cod populations can ever be established in lowland habitats, and therefore the issue of upland habitats, and the return of some upland habitats in trout-free form, needs to be addressed in trout cod conservation.

Historical evidence indicates trout cod (and Macquarie perch) were abundant in most of the larger upland rivers and streams in the southeast corner of the Murray-Darling river system, and that these upland river habitats were their primary habitats. The extinction of trout cod populations in every one of its upland river habitats is an unresolved issue. Contrary to popular belief, many of these upland rivers still contain significant stretches of unregulated, high quality habitat.

While dams, thermal pollution, siltation and other forms of habitat modification and degradation are responsible for the trout cod's extinction in many upland river habitats, it is almost certain that the reason for the trout cod's extinction in higher quality upland river habitats, that have not experienced serious modification and degradation, is the heavy domination of these habitats by introduced trout species, which are aggressive, predatory fish.

Every single larger upland river and stream in southeastern Australia is dominated by introduced trout species (rainbow trout and brown trout), with many having been continually stocked with introduced trout species for more than a century, and not a single larger upland river or stream in southeastern Australia has been reserved in a trout-free state for larger upland native fish species. The effects of this course of action by fishery departments has been stark, and catastrophic events such as drought or bushfire, after which introduced trout species were restocked but upland native fish were not restocked, have shifted the balance further. The net result is that a number of upland native fish species including trout cod (and Macquarie perch) have completely died out or nearly so in their upland river habitats in the wild, apparently unable to cope with massive predation on their larvae/juveniles by introduced trout species and unable to cope with massive competition from introduced trout species for food and habitat at all life stages.

Two small populations of trout cod that have shown indications of breeding have been created by hatchery stockings in the lowland reaches of Murrumbidgee River at Gundagai and Narranderra, although it is far from clear whether these populations will be self-sustaining in the long term. Many other stockings of trout cod have failed, which is not surprisingly considering the small number of trout cod fingerlings stocked, and the that fact that trout cod were frequently stocked into upland river habitats where introduced trout species were heavily entrenched, and in at least one case, were carried in conjunction with far larger stockings of introduced trout. A semi-natural population exists in a stretch of the very small upland Seven Creeks, which was established by translocations of trout cod (and Macquarie perch) above a set of falls in the 1920s. The Seven Creeks population is not a wholly-artificial population, or a wholly unrepresentative habitat, as often claimed, as some of the trout cod translocated came — literally — from the base of the falls.

The future

With emerging historical evidence emphasising upland river habitats as the original primary habitat for trout cod (and

Macquarie perch), and casting serious doubt over attempts to re-establish self-sustaining populations in lowland habitats, this species will likely remain endangered until it is provided with a few sizeable, trout-free upland river habitats.

This process could be undertaken with no serious impacts on the economically and recreationally important introduced trout fisheries in Australia, and indeed could be of great benefit to fishermen and fisheries. However, this appears to be a nettle that no fishery or conservation agency is prepared to grasp, again due to strong cultural cringe towards introduced trout in the wider community and the management bias towards introduced trout amongst fishery agencies.

In the meantime the fishers of southeastern Australia are missing out on a unique part of their heritage; a beautiful upland native fish that is vastly superior in sporting and fighting qualities compared to introduced trout species, and was revered by early fishers.

Trumpetfish

Trumpetfish, *Aulostomus maculatus*, are long bodied fish that often swim vertically while trying to blend with vertical coral, like sea rods, sea pens, and pipe sponges.

Habitat

Trumpetfish occur in waters between 3 and 30 metres deep and can grow to 40 to 80 cm in length. They are sometimes locally abundant over coral atoll reefs or in lagoons, where they may be caught even in areas of severe wave action. The spawning habits of the trumpetfish are unknown, but in the region around Madeira, it is known that the females have mature eggs from March to June.

Description

Trumpetfish are closely related to cornetfish. Trumpetfish are less than 2 feet (60 cm) long and have greatly elongated bodies with small jaws at the front end of a long, tubular snout. The gills are pectinate, resembling the teeth of a comb, and a soft dorsal fin is found near the tail fin. A series of spines occurs in front of the dorsal fin. Trumpetfish vary in colour from dark brown to greenish but also yellow in some areas. A black streak, sometimes reduced to a dark spot, occurs along the jaw, and a pair of dark spots is sometimes found on the base of the tail fin.

Trumpetfish swim slowly, sneaking up on unsuspecting prey, or lie motionless like a floating stick, swaying back and forth with the wave action of the water. They are adept at camouflaging themselves and often swim in alignment with other larger fishes. They feed almost exclusively on small cardinal fishes, such as wrasses and silversides.

Trumpetfish make up the genus *Aulostomus* of the family Aulostomidae.

Tube-eye

The tube-eye or thread-tail, *Stylephorus chordatus*, is a deep-sea Lampriformes fish, the only fish in the genus *Stylephorus* and family Stylephoridae.

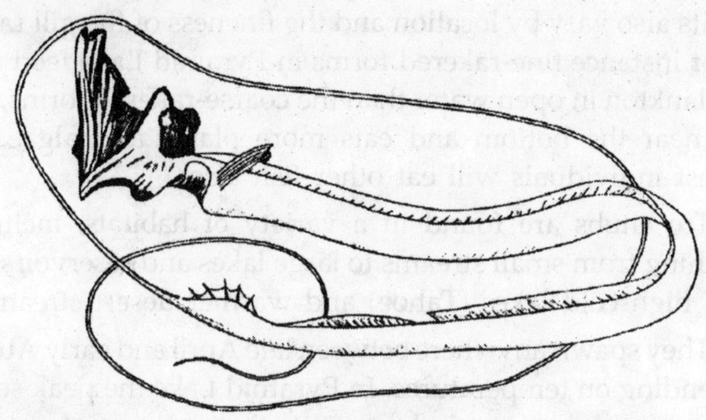

It is found in deep subtropical and tropical waters around the world, living at depths during the day and making nightly vertical migrations to feed on plankton. It is an extremely elongated fish: although its body grows only to 28 cm long, it has a pair of tail fin rays that triple its length to about 90 cm. Its eyes bulge out from the head on telescopic tubes.

It has a tubular mouth through which it sucks seawater by enlarging its oral cavity to about forty times its original size. It then expels the water through the gills, leaving behind the copepods on which it feeds.

Tui Chub

The tui chub *Gila bicolor* is a cyprinid fish native to western North America. Widespread in many areas, it is an important food source for other fish, including the cutthroat trout.

The form and appearance of the tui chub is variable; many were originally described as different species by J. O. Snyder, but have since been reduced to subspecies. In general, the colour is deep olive above and white below, with a smooth variation in shading along the sides, and a brassy reflection. Fins are olive and sometimes tinted with red. The pectoral fins are far forward and low on the body. Length has been recorded at up to 45 cm, but 25 cm is more typical.

Tui chub diet is varied; young fish eat mostly invertebrates, adding plant material and especially algae as they mature. Habits also vary by location and the fineness of the gill rakers, so for instance fine-rakered forms in Pyramid Lake feed more on plankton in open water than the coarse-rakered forms, who live near the bottom and eats more plants and algae. The largest individuals will eat other fish also.

Tui chubs are found in a variety of habitats, including anything from small streams to large lakes and reservoirs, and both high cold lakes (Tahoe) and warmer desert streams.

They spawn anywhere between late April and early August, depending on temperatures. In Pyramid Lake the peak season is June; males move inshore first, then congregate around arriving females in shallow water, preferring areas of heavy vegetation. The female scatters her eggs randomly over a wide area, where they are then fertilized by several males. The hatchlings remain in the heavy vegetation for the remainder of the summer. In Lake Tahoe some chubs spawn around stream mouths in July.

The tui chub's range includes the Lahontan and Central system of the Great Basin, as well as the Owens and Mojave Rivers. It is found in the Pit River and Goose Lake of the upper Central Valley, in the Klamath River system, and in the Columbia River drainage.

Subspecies

The exact number of subspecies is not known; Sigler & Sigler estimate as high as 16. Agreed subspecies include:

- *Gila bicolor bicolor*
- *Gila bicolor isolata*
- *Gila bicolor mohavensis*
- *Gila bicolor obesa*
- *Gila bicolor pectinifer*
- *Gila bicolor snyderi*

Tuna

Tuna are several species of ocean-dwelling fish in the family Scombridae, mostly in the genus *Thunnus*. Some tuna are able to inhabit freshwater environs as well. Tunas are fast swimmers—have been clocked at 77 km/h (48 mph)—and include several species that are warm-blooded. Unlike most fish species, which have white flesh, the flesh of tuna is pink to dark red. This is because tuna muscle tissue contains greater quantities of myoglobin, an oxygen-binding molecule.

Some of the larger tuna species such as the bluefin tuna can raise their blood temperature above the water temperature with muscular activity. This enables them to live in cooler waters and survive a wider range of circumstances. Some tuna species and fisheries are overfished and there are risks of some tuna fisheries collapsing.

Commercial Importance

Tuna is an important commercial fish. Some varieties of tuna, such as the bluefin and bigeye tuna, *Thunnus obesus*, are threatened by overfishing, dramatically affecting tuna populations in the Atlantic and northwestern Pacific Oceans. Other populations seem to support fairly healthy fisheries (for example, the central and western Pacific skipjack tuna, *Katsuwonus pelamis*), but there is mounting evidence that overcapitalisation threatens tuna fisheries worldwide.

The Australian Government alleged in 2006 that Japan had illegally overfished southern bluefin to the value of USD $2 billion, by taking 12,000 to 20,000 tons per year instead of the internationally agreed 6,000 tons.

This has resulted in severe damage to stocks. "Japan's huge appetite for tuna will take the most sought-after stocks to the brink of commercial extinction unless fisheries agree on more rigid quotas, wildlife campaigners warned today" stated by the WWF. Some say this is unfortunately in accord with the Japanese government's refusal to deal with sustainable environmental ideas and the sometime utter neglect of animal rights or preservation needs, such as fishing quotas.

Increasing quantities of high-grade tuna are entering the market from operations that rear tuna in net pens and feed them on a variety of bait fish. In Australia the southern bluefin tuna, *Thunnus maccoyii,* is one of two species of bluefin tunas that are kept in tuna farms by former fishermen. Its close relative, the northern bluefin tuna, *Thunnus thynnus,* is being used to develop tuna farming industries in the Mediterranean, North America and Japan.

Due to their high position in the food chain and the subsequent accumulation of heavy metals from their diet, mercury levels can be relatively high in some of the larger species of tuna such as bluefin and albacore. As a result, in March 2004 the United States FDA issued guidelines recommending pregnant women, nursing mothers and children limit their intake of tuna and other types of predatory fish . However, most canned light tuna is skipjack tuna and is lower in mercury. The Chicago Tribune reported that some canned light tuna such as yellowfin tuna is significantly higher in mercury than skipjack tuna, and caused Consumer Reports and other health groups to advise pregnant women to refrain from consuming canned tuna. Further, the Eastern Little Tuna (*Euthynnus affinis*) has been available for decades as a low-mercury, less expensive canned tuna. Of the five major species of canned tuna imported into the United States, however, it is the least commercially attractive, mostly due to its dark colour and more pronounced 'fishy' flavour. Its use has traditionally been exclusively restricted to institutional (non-retail) commerce.

Canned Tuna

Canned tuna was first produced in 1903, and quickly became popular. In the United States, only Albacore can legally be sold in canned form as "white meat tuna"; in other countries, Yellowfin is also acceptable as "white meat tuna."

While in the early 80s canned tuna in Australia was most likely to be Southern bluefin, today (as of 2003) it is usually yellowfin, skipjack, or tongol (labelled "northern bluefin")..

As tuna are often caught long distances from where they're processed, poor quality control may lead to spoilage. The canning process kills any bacteria, but retains the histamine produced by the fish and rancid flavours. The international standard sets the maximum histamine level at 200 mg per kg. An Australian study of 53 varieties of unflavoured canned tuna found none to exceed the maximum histamine level, although some had "off" flavours.

Standards once required canned tuna to contain at least 51 per cent tuna in Australia, but these regulations were dropped in 2003.

Major Tuna Producers

According to Foodmarket Exchange, total tuna catching stood at 3,605,000 tons in 2000, down about 5.7 per cent from 3,823,000 tons in 1999. The main tuna catching nations are concentrated in Asia, with Japan and Taiwan as the main producers. Other important tuna catching nations in Asia are Indonesia and South Korea.

Spain and France are important tuna fishing countries, mainly catching in the Indian Ocean.

Japan remains the main nation fishing for tuna in the Pacific. In 2000, total tuna caught by Japanese vessels stood at 633,000 tons, about 17 per cent of the world tuna catch. Taiwan was the second biggest tuna producer at 435,000 tons, or about 12 per cent of the total tuna catch. Spain supplies most of the yellowfin to European canneries, accounting for 5.9 per cent of the total tuna catch, while Ecuador and Mexico dominate the Eastern Pacific Ocean.

Management and Conservation

There are 5 main tuna fishery management bodies. The five are the Western Central Pacific Ocean Fisheries Commission, the Inter-American Tropical Tuna Commission, the Indian Ocean Tuna Commission, the International Commission for the Conservation of Atlantic Tunas and the Commission for the Conservation of Southern Bluefin Tuna. They met for the first time in Kobe in Japan in January 2007. Environmental organisations made submissions on risks to fisheries and species.

The meeting concluded with an action plan drafted by some 60 countries or areas. Concrete steps include issuing certificates of origin to prevent illegal fishing and greater transparency in the setting of regional fishing quotas. The delegates will meet again at the second joint meeting in January or February 2009 in Europe.

Association with Dolphins

Many tuna species associate with dolphins, swimming alongside them. These include yellowfin tuna in the eastern Pacific Ocean. Species which do not include albacore and skipjack. The reason for the association is believed to be the avoidance of dolphins by sharks, which are predators of tuna. Swimming near dolphins reduces the likelihood of the tuna being attacked by a shark. Methods of fishing tuna have become more "dolphin friendly", becoming less prone to entangle, injure or kill dolphins. However, there are no universal independent inspection programme or verification of "dolphin safeness" to show that dolphins are not harmed during tuna fishing. According to the Consumers Union, this gives the claims such as "dolphin safe" little credibility.

Methods of Capture

- Arabic method of Almadraba, in which nets are used, creating a maze in which the tuna are secured.
- big-game fishing
- longline fishing
- purse seine - the major threat to dolphins
- pole-and-line

Species

There are eight tuna species in the *Thunnus* genus:

- Albacore, *Thunnus alalunga* (Bonnaterre, 1788).
- Yellowfin tuna, *Thunnus albacares* (Bonnaterre, 1788).
- Blackfin tuna, *Thunnus atlanticus* (Lesson, 1831).
- Southern bluefin tuna, *Thunnus maccoyii* (Castelnau, 1872).
- Bigeye tuna, *Thunnus obesus* (Lowe, 1839).
- Pacific bluefin tuna, *Thunnus orientalis* (Temminck & Schlegel, 1844).
- Northern bluefin tuna, *Thunnus thynnus* (Linnaeus, 1758).
- Longtail tuna, *Thunnus tonggol* (Bleeker, 1851).

Species of several other genera (all in the family Scombridae) have common names containing "tuna":

- Skipjack tuna *Katsuwonus pelamis* (Linnaeus, 1758).
- Slender tuna *Allothunnus fallai* (Serventy, 1948).
- Bullet tuna *Auxis rochei rochei*
- Frigate tuna *Auxis thazard thazard*
- Kawakawa (little tunafish or mackerel tunafish) *Euthynnus affinis* (Cantor, 1849).

- Little tunny (little tunafish) *Euthynnus alletteratus*
- Butterfly kingfish (Butterfly mackerel) *Gasterochisma melampus* (Richardson, 1845)
- Dogtooth tuna *Gymnosarda unicolor* (Ruppell, 1836)

Turbot

Turbot [pronounced *tur-bit* or alternately *tur-bo*] (family Scophthalmidae, order Pleuronectiformes) are flatfish native to marine or brackish waters of the North Atlantic. The taxon name comes from the Greek language, with *skopein* meaning "to look" and *ophthalmos* meaning "eye".

The European turbot (*Psetta maxima*) is a large left-eyed flatfish found primarily close to shore in sandy shallow waters throughout the Mediterranean, the Baltic Sea, the Black Sea and the North Atlantic. The European turbot has an asymmetrical disk-shaped body, and may attain sizes of 30 to 40 pounds (approx. 15 to 17 kilograms).

Turbot is highly prized for its delicate flavour, and is a valuable commercial species, acquired through aquaculture and trawling. Turbot are farmed in France, Spain, Chile, Norway and China.

Turbot has a bright white flesh that retains its appearance when cooked. Like all flatfish, turbot yields four fillets with meatier topside portions that may be baked, poached or pan-fried. American English speakers often pronounce turbot as *tur-bo*, likely a back-formation based on French words ending in *-ot*, and it is pronounced as such in the French language as well.

Greenland Turbot, sometimes known as blue halibut, is usually harvested in the cold waters off Greenland in water depths of up to 1000m. Both flavour and texture is very similar to Pacific Halibut.

Dwarf Upside-down Catfish

Upside-down catfish, *Synodontis nigriventris*, is a species of catfish. It's particularly noteworthy because of its habit of swimming upside down most of the time. Upside-down catfish originate from the Central Congo basin of Africa.

Appearance and Anatomy

Upside-down catfish are small, reaching a maximum of 9.6 centimetres (4 in). Upside-down catfish are adapted to spending most of their time upside-down. This is reflected in the fish's pigmentation—their bellies are darker than their backs, a form of countershading.

Ecology

These fish are mostly nocturnal, and feed on insects, crustaceans, and plant matter. These fish lay eggs. The young fish do not swim upside-down until they are about two months old.

In the Aquarium

The upside-down catfish is well suited to aquariums because of its small size (typically 9 or 10 cm or less) and peaceful demeanour. It fares best in schools of up to half a dozen.

Viperfish

A Viperfish (genus *Chauliodus*) is a predatory fish of the deep abyssal plain. This fish can be recognised by its large mouth and sharp, fang-like teeth. These fangs are so large in fact that they do not fit inside its mouth. Instead, they curve back very close to the fish's eyes. The viper is thought to use these sharp teeth to impale its victims by swimming at them at high speeds. The first vertebra, right behind the head, actually acts as a shock absorber.

This fearsome looking creature has a long dorsal spine that is tipped with a photophore, a light-producing organ. The viperfish uses this light organ to attract its prey. By flashing it on and off, it can be used like a fishing lure to attract smaller fish. They have been known to hang motionless in the water, waving their lures over their heads to attract their meals. Vipers have a hinged skull, which can be rotated up for swallowing large prey. They also have large stomachs that allow them to stock up on food whenever it is plentiful. The viperfish also has photophores all along the sides of its body. These light organs may be used to signal and attract other viperfish during mating. Like many deep-sea creatures, the viperfish is known to migrate vertically throughout the day. During daytime hours they are found in deep water down to 5000 feet. At night they travel up onto shallower waters at depths of less than 2000 feet where food is more plentiful. The viperfish grows to between 12 and 24 inches in length and is found in most waters of the world.

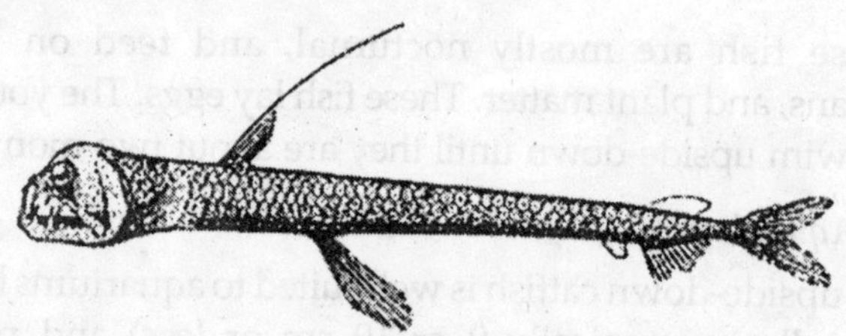

The name viperfish is also sometimes applied to the lesser weever.

Species

There are nine species:

- *Chauliodus barbatus* (Lowe, 1843).
- Dana viperfish, *Chauliodus danae* (Regan & Trewavas, 1929).
- *Chauliodus dentatus* (Garman, 1899).
- Pacific viperfish, *Chauliodus macouni* (Bean, 1890).
- *Chauliodus minimus* (Parin & Novikova, 1974).
- *Chauliodus pammelas* (Alcock, 1892).
- *Chauliodus schmidti* (Regan & Trewavas, 1929).

Wahoo

The Wahoo (*Acanthocybium solandri*) is a dark blue scombrid fish found worldwide in tropical and subtropical seas. Some say that the name "Wahoo" is a derivation of the name of the Hawaiian Island Oahu, while others say the name derives from the exclamation of some fishermen, "Wahoo!" who have hooked into the extremely fast running fish. The fish is also known as Ono, after the Hawaiian word for "delicious", *'ono,* Jack Mackerel, and Peto.

The body is elongate and covered with small, scarcely visible, scales; the back is an iridescent blue green, while the sides are silvery, with a pattern of vertical blue bars. These colours fade rapidly at death. The mouth is large, and both the upper and lower jaws have a somewhat sharper appearance than those of King or Spanish Mackerel. Specimens have been recorded at up to 2.5 metres (8 ft) in length, and weighing up to 83 kilograms (180 lb). Growth can be rapid. One specimen tagged at 11 pounds grew to 33 pounds in one year. Wahoo can swim up to 75 kilometres (47 miles) per hour.

The Wahoo may be distinguished from the related King mackerel by a fold of skin which covers the mandible when its mouth is closed. In contrast, the mandible of the King mackerel is always visible as is also the case for Spanish and Cero mackerels. Their teeth are similar to those of King mackerel, but shorter and more closely set together.

Wahoos tend to be solitary or occur in loose-knit groups of two or three fish, rather than in schools. Their diet consists essentially of other fish and squid.

The flesh of the Wahoo is delicate and white and regarded as very good in quality. This has created some demand for the wahoo as a premium priced commercial food fish. However, because of its solitary lifestyle, it is taken commercially only as a bycatch in the long-line fishery for Tuna and Dolphin. Wherever found, it is a prized sport fishing catch. Not always catchable, but otherwise it can be very delicious.

Most wahoo taken have a trematode parasite *(Hirudinella ventricosa)* living in their stomach. It appears to do no harm to the fish.

Walking Catfish

The walking catfish, *Clarias batrachus,* which is popularly known in malay (Malaysia and Indonesia respectively) as ikan keli or ikan sembilang. It is also known as the magur or pla duk dam to another region, is a species of airbreathing catfish with the ability to "walk" out of the water and across land.

Its "walk" is more like a sort of wriggling motion with snakelike movements, as well as using its pectoral fins as "legs". This fish normally lives in slow-moving and often stagnant waters in ponds, swamps, streams and rivers (Mekong and Chao Phraya basins), flooded rice paddies or temporary pools which may dry up. When this happens, its "walking" skill comes in handy for moving to other sources of water. In Tamil it is known as *keluthi* or *keluthu*.

In the wild, the natural diet of this creature is omnivorous; it feeds on smaller fish, molluscs and other invertebrates as well as detritus and aquatic weeds. It is a voracious eater which consumes food rapidly and this habit makes it a particularly harmful invasive species.

The walking catfish is a native of South East Asia including Malaysia, Thailand, eastern India, Sri Lanka, Bangladesh, Burma, Indonesia, Singapore, and Borneo. It was probably introduced into the Philippines. The catfish is a tropical animal and prefers a water temperature in the range of 10-28°C.

In the United States it is a non-indigenous invasive species, which is now established in Florida and reported from California, Connecticut, Georgia, Massachusetts, and Nevada.

The walking catfish was imported to Florida, reportedly from Thailand, in the early 1960s for the aquarium trade. The first introductions apparently occurred in the mid-1960s when adult fish imported as brood stock escaped, either from a fish farm in northeastern Broward County or from a truck transporting brood fish between Dade and Broward counties. Additional introductions in Florida, supposedly purposeful releases, were made by fish farmers in the Tampa Bay area, Hillsborough County in late 1967 or early 1968, after the state banned the importation and possession of walking catfish. Aquarium releases likely are responsible for introductions in other states. Dill and Cordone (1997) reported that this species has been sold by tropical fish dealers in California for some time.

In Florida, walking catfish are known to have invaded

aquaculture farms, entering ponds where these predators prey on fish stocks. In response, fish farmers have had to erect protective fences to protect ponds. Authorities have also created laws that banned possession of Walking Catfishes.

This fish needs to be handled carefully when fishing it out due to its hidden embedded sting or thorn like defensive mechanism hidden behind its fins (including the middle ones before the tail fin).

Wallago

Wallago is a genus of catfish in the family Siluridae, or "sheatfishes". Although the genus has 5 recognised species, the name "wallago" is also used as a common name for *Wallago attu*.

Species

- *Wallago attu*
- *Wallago hexanema*
- *Wallago leerii*
- *Wallago maculatus*
- *Wallago micropogon*

Walleye

The Walleye (*Sander vitreus vitreus,* formerly *Stizostedion vitreum vitreum*) is a freshwater perciform fish native to most of Canada and to the northern United States. It is a North American close relative of the European pikeperch. The walleye is sometimes also called the yellow walleye to distinguish it from the extinct blue walleye.

In some parts of its range, the walleye is also known as the walleyed pike, yellow pike or pickerel (esp. in English-speaking Canada), although the fish is related neither to the pikes nor to the pickerels, both of which are members of the family Esocidae.

Genetically, walleyes show a fair amount of variation across watersheds. In general, fish within a watershed are quite similar and are genetically distinct from those of nearby watersheds.

The species has been artificially propagated for over a century and has been planted on top of existing populations or introduced into waters naturally devoid of the species, sometimes reducing the overall genetic distinctiveness of populations.

Meaning of the Name

The common name, "walleye", comes from the fact that their eyes, like those of cats, reflect light. This is the result of a light-gathering layer in the eyes called the *tapetum lucidum* which allows the fish to see well in low-light conditions. In fact, many anglers look for walleyes at night since this is when most major feeding patterns occur.

Their eyes also allow them to see well in turbid waters (stained or rough, breaking waters) which gives them an advantage over their prey.

Thus, walleye anglers will commonly look for days and locations where there is a good "walleye chop" (*i.e.* rough water). This excellent vision also allows the fish to populate the deeper regions in a lake and can often be found in deeper water

Physical Description

Walleyes grow to about 75 cm (30 in) in length, and weigh up to about 7 kg (15 lb). The maximum recorded size for the fish is 107 cm (42 in) in length and 11.3 kg (25 lb) in weight. The growth rate depends partly on where in their range they occur, with southern populations often growing faster and larger. In general, females grow larger than males.

Walleyes may live for decades; the maximum recorded age is 29 years. In heavily fished populations, however, few walleye older than 5 or 6 years of age are encountered.

Walleyes are largely olive and gold in colour (hence the French common name: *dore*—golden). The dorsal side of a walleye is olive, grading into a golden hue on the flanks. The olive/gold pattern is broken up by five darker saddles that extend to the upper sides. The colour shades to white on the belly.

The mouth of a walleye is large and is armed with many sharp teeth. The first dorsal and anal fins are spinous as is the operculum. Walleyes are distinguished from their close cousin the sauger by the white colouration on the lower lobe of the caudal fin which is absent on the sauger. In addition, the two dorsals and the caudal fin of the sauger are marked with distinctive rows of black dots which are absent from or indistinct on the same fins of walleyes.

Habitat

The walleye is a relatively selected species. They require fairly clean waters and are found most often in deep mesotrophic lakes and moderate- to low-gradient rivers. The walleye is considered a "cool water" species.

Reproduction

In most of the species' range, the majority of male walleyes mature at age 3 or 4. Females normally mature about a year later. Adults migrate to tributary streams in late winter or early spring to lay eggs over gravel and rock, although there are open water reef or shoal spawning strains as well.

Some populations are known to spawn on sand or on vegetation. Spawning occurs at water temperatures of 6 to 10° C (43 to 50° F). A large female can lay up to 500,000 eggs and no care is given by the parents to the eggs or fry. The eggs are slightly adhesive and fall into spaces between rocks. The incubation period for the embryos is temperature-dependent but generally lasts from 12 to 30 days.

After hatching, the free-swimming embryo spends about a week absorbing the relatively small amount of yolk. Once the yolk has been fully absorbed, the young walleye begins to feed on invertebrates such as fly larvae and zooplankton. After 40 to 60 days, juvenile walleyes become piscivorous. Thenceforth, both juvenile and adult walleyes eat fish almost exclusively, frequently yellow perch or ciscoes, moving onto bars and shoals at night to feed. Walleye also feed heavily on crayfish, minnows, leeches, and earthworms.

Interaction with Humans

The walleye is often considered to have the best tasting flesh of any freshwater fish, and, consequently, is fished recreationally and commercially. Because of its nocturnal feeding habits, it is most easily caught at night using live minnows or lures that mimic small fish. Most commercial fisheries for walleye are situated in the Canadian waters of the Great Lakes, but there are other locations as well.

Fishing

The walleye is a light-avoiding fish, caught most often under low light conditions. Fishing is generally best on cloudy or overcast days, or on days when waves keep light from penetrating too deeply into the water.

"Walleye chop" is a term used by walleye anglers for rough water typical with winds of 5 to 15 mph (7 to 24 km/h), and is one of the indicators for good walleye fishing due to the walleye's increased feeding activity during such conditions. The eyes of a walleye allow it to see much better than their prey in breaking waters.

Because walleyes are popular with anglers, fishing for walleyes is regulated by most natural resource agencies. Management may include the use of quotas and length limits to ensure that populations are not overexploited.

Seasons

In springtime walleye will take almost any bait or lure, but may be more challenging to catch through the summer months. Fall often brings another peak of walleye feeding activity. Walleye are readily caught through the ice in winter, usually on jigs, jigging spoons or minnows.

Bait

Casting or trolling with spinners or minnow-imitating plugs is a good bet. Special worm harness rigs of spinners and beads are often trolled. Jigs, either traditional bucktails, or tipped with any of the modern plastics, a piece of worm or minnow are walleye angling favourites.

Live baits are often still-fished, drifted or trolled on slip-sinker or "bottom-bouncing" rigs. Excellent live bait includes minnows, earthworms and crayfish.

When ice fishing walleye are caught jigging or on tip-ups. Tip-ups are generally set up with a dacron backing and a clear synthetic leader. For bait, the most common minnows are suckers and shiners. Size for bait is anywhere from 1 to 7 inches.

Minnesota

The walleye is the state fish of Minnesota. Its popularity with Minnesota residents means that the residents of that state consume more of the fish than in any other jurisdiction. In 2004, it was revealed that some restaurants in the Minneapolis-St. Paul region had been substituting the less expensive, imported zander for the walleye indicated on the menu.

Zander (pikeperch) is a closely-related species and is almost impossible to tell apart by taste, so the television station that did the expose had to send samples of food for DNA testing. Though sold as "walleye", several samples were found to be zander, which is considered an illegal practice by the US Food and Drug Administration.

Columbia River

Walleye are an introduced species in the Columbia River. Given their pisiverous nature, they likely prey on juvenile salmon and steelhead some of which are listed under the US Endangered Species Act. Walleye support a popular sport fishery in the Columbia and are sold commercially in small numbers by tribal fishers.

Warmouth

The warmouth (*Lepomis gulosus*) is a species of freshwater fish. It is a member of the sunfish family (family Centrarchidae) of order Perciformes. It is native to a wide area of the United States, from Minnesota to western Pennsylvania in the north and from the Rio Grande in New Mexico east to the Atlantic in the south, inhabiting the heavily vegetated, muddy-bottomed habitats typical of the sunfishes.

The warmouth's classification is disputed. Cuvier, in 1829, originally described it as a member of the obsolete genus *Pomotis,* and it was reclassified much later as a member of *Chaenobryttus* Gill, 1864, on the basis of genetic evidence. ITIS retains this classification, with the warmouth the only member of that genus. FishBase, however, follows the majority of the sources in considering the warmouth to be a member of *Lepomis, Chaenobryttus* being a subgenus thereof.

The warmouth's specific epithet, *gulosus,* derives from the Latin *gulosus* (gluttonous). The origin of the common name "warmouth" is due to the stripes around the mouth of the fish bearing a resemblance to warpaint.

Besides their distinctive spotted patterns, warmouths can be easily identified and distinguished from other panfish or sunfish by their proportionally large mouths. Other panfish of the same weight will have mouths 2-3 times smaller than that of the warmouth, who's oral proportions are in line with those of largemouth/black bass. It is often stated that warmouths have the body of a bream (a small common sunfish) and the head of a bass, leading to rumours that it is a man-made crossbreed.

Warmouths can be very aggressive, and will often strike at lures and baits even after being released by anglers moments before.

4

Notosudidae

Notosudidae have a circumglobal distribution in sub-Arctic to sub-Antarctic waters, the family contains just 17 species in 3 genera. The family name Notosudidae derives from the Greek noton (back) and Latin sudis (a fish, esox, the name of salmon).

Physical Description

Dorsal fin 9-14 rays, anal fin 16-21 rays, pectoral fin 10-15 rays. Scales in lateral line 44-65. The swim bladder is absent, as are any photophores.

Species

- Genus *Ahliesaurus*
  - — *Ahliesaurus berryi* (Krefft & Marshall, 1976).
  - — *Ahliesaurus brevis* (Krefft & Marshall, 1976).
- Genus *Luciosudis*
  - — *Luciosudis normani* (Fraser-Brunner, 1931).
- Genus *Scopelosaurus*
  - — *Scopelosaurus adleri* (Bertelsen, Krefft & Marshall, 1976).
  - — *Scopelosaurus ahlstromi* (Bertelsen, Krefft & Marshall, 1976).

— *Scopelosaurus argenteus* (Maul, 1954).

— *Scopelosaurus craddocki* (Bertelsen, Krefft & Marshall, 1976).

— *Scopelosaurus gibbsi* (Bertelsen, Krefft & Marshall, 1976).

— *Scopelosaurus hamiltoni* (Waite, 1916).

— *Scopelosaurus harryi* (Mead, 1953).

— *Scopelosaurus herwigi* (Bertelsen, Krefft & Marshall, 1976).

— *Scopelosaurus hoedti* (Bleeker, 1860).

— *Scopelosaurus hubbsi* (Bertelsen, Krefft & Marshall, 1976).

— *Scopelosaurus lepidus* (Krefft & Maul, 1955).

— *Scopelosaurus mauli* (Bertelsen, Krefft & Marshall, 1976).

— *Scopelosaurus meadi* (Bertelsen, Krefft & Marshall, 1976).

— *Scopelosaurus smithii* (Bean, 1925).

Wels Catfish

The wels catfish, *Silurus glanis,* is a scaleless freshwater catfish recognisable by its broad, flat head and wide mouth. The mouth contains lines of numerous small teeth, two long barbels on the upper jaw and four shorter barbels on the lower jaw. It has a long anal fin that extends to the caudal fin, and a small sharp dorsal fin positioned relatively far forward. It uses its sharp pectoral fins to capture prey: with these fins, it creates an eddy to disorient its victim, which it then simply engulfs in its enormous throat. It has very slippery green-brown skin. Its belly is pale yellow or white. Wels catfish can live for at least thirty years and have very good hearing.

The female produces up to 30,000 eggs per kilogram of body weight. The male guards the nest until the brood hatches, which, depending on water temperature, can take from three to ten days. If the water level decreases too much or too fast the male has been observed to splash the eggs with the muscular tail in order to keep them wet.

The wels catfish lives on annelid worms, gastropods, insects, crustaceans, and fish; the larger ones also eat frogs, mice, rats and aquatic birds such as ducks.

Habitat

The wels catfish lives in large, warm lakes and deep, slow-flowing rivers. It prefers to remain in sheltered locations such as holes in the riverbed, sunken trees, etc. It consumes its food in the open water or on the bottom, where it can be recognised by its large mouth. Wels catfish are food fish and are also kept in fish ponds. However, only the meat of younger animals is palatable.

The wels catfish is found in wide areas of central, southern, and eastern Europe, and near the Baltic and Caspian Seas.

Physical Characteristics

Colour varies with environment : clear water will give the fish a black colouration while muddy water will often tend to produce brownish specimens. Weight and length are not correlated linearly, and also depends on the season.

With a possible total length up to 3 m (ten feet) and a maximum weight of over 150 kg (330 lbs) it is the second largest freshwater fish in its region after the sturgeon. However, such lengths are extremely rare and could not be proved during the last century, but there is a somewhat credible report from the 19th century of a wels catfish of this size.

Most wels catfish are only about 1.3 to 1.6 m (4 ft 3 in to 5 ft 3 in) long; fish longer than 2 m (6 ft 6 in) are normally extremely rare. At 1.5 m they can weigh 15 to 20 kg and at 2.20 m they can weigh 65 kg.

Only under exceptionally good living circumstances can the wels catfish reach lengths of more than 2 m, as with the record wels catfish of Kiebingen (near Rottenburg, Germany), which was 2.49 m (8 ft 2 in) long and weighed 89 kg (196 lb). This giant was surpassed by some even larger specimens from Poland, France, Spain (in the River Ebro), Italy (in the River Po), and Greece, where this fish was released a few decades ago.

It grows very well at that location thanks to the mild climate, lack of competition, and good food supply. The record sized was 3 m (9.9 ft) long fish, caught in the Danube in Romania, weighing about 220 kg (440 lb), until only recently when a 226lb (102kg), 8ft (2.4m) monster was caught in the Ebro River in Spain by Carl Smith. Other reports of larger wels are unlikely and are often regarded as typical big fish stories or in some cases misidentifications of the now rare sturgeon.

Ecology

There are concerns about the ecological impact of introducing the wels catfish to non-native regions. These concerns take into account the situation in Lake Victoria in Africa, where Nile perch (available in stores as *Lake Victoria perch*) were introduced and rapidly caused the extinction of numerous indigenous species.

This severely impacted the entire lake, destroying much of the original ecosystem. The danger does not appear to be as extreme in the case of the wels catfish, but the introduction of foreign species is almost always a burden on the affected ecosystem.

As Food Fish257

Meat of *Silurus glanis* is regarded as good to eat when the catfish has a weight under 15 kg. Larger than this size, the fish is highly fatty and not recommended for consumption. It should be noted that the eggs are poisonous and should not be consumed. In sign of respect for the strong defence during the fishing phase, often fisherman release it (catch and release philosophy).

Fishing

For fishing techniques, one of the best baits are the bloodsuckers, turned inside-out on the hooks and one bait that nearly never fails and is guaranteed to attract one catfish is the fresh water eel. Attached carefully to the hook, not to break the spine, it will remain alive for a long period of time while emitting sounds that attract the catfish. Other common baits are different types of dead and living baitfish, clumps of large earthworms and sometimes also innards or pieces of squid. During the winter is fished with a special trawler that sweeps the river bed. Once caught, the fish is very slippery and powerful and difficult to master in the boat. Once there, is a quiet, no dangerous fish. While some specimens clearly reach lengths over 2.50 m, there are no recorded attacks on humans.

Related Species

- Aristotle catfish (*Silurus aristotelis*) from Greece, the only other native European catfish species beside *Silurus glanis*.
- Amur catfish (*Silurus asotus*), introduced to European rivers.
- Brown bullhead (*Ameiurus nebulosus*).
- Giant Mekong catfish on the lower reaches of the Mekong.

Whitebait

Whitebait are young fish; in Europe the term applies to young herring, but in other parts of the world it is used for similar fish of other species. Whitebait are tender and edible. The entire fish is eaten including head, fins and gut but typically each 'bait' is only 25-50 mm in length and about 3 mm in cross-section. Whitebaiting is the activity of catching whitebait.

New Zealand Whitebait

New Zealand whitebatty are the juvenile of certain galaxiids which mature and live as adults in rivers with native forest surrounds. The larvae of these galaxiids is swept down to the ocean where they hatch and the sprats then move back up their home rivers as whitebait.

The most common whitebait species in New Zealand is the common galaxias or inanga, which lays its eggs during spring tides in Autumn on the banks of a river amongst grasses that are flooded by the tide. The next spring tide causes the eggs to hatch into larvae which are then flushed down to the sea with the outgoing tide where they form part of the ocean's plankton mass. After six months the developed juveniles return to rivers and move upstream to live in freshwater.

New Zealand whitebait are caught in the lower reaches of the rivers using small open-mouthed hand-held nets although in some parts of the country where the whitebait is more plentiful, larger (but not very large) set nets may be used adjacent to river banks. Whitebaiters constantly attend the nets in order to lift them as soon as a shoal enters the net. Otherwise. the whitebait quickly swim back out of the net. Typically, the small nets have a long pole attached so that the whitebaiter can stand on the river bank and scoop the net forward and out of the water when whitebait are seen to enter it. The larger nets may be set into a platform extending into the river from the bank and various forms of apparatus used to lift the net.

Whitebaiting in New Zealand is a seasonal activity with a fixed and limited period enforced during the period that the whitebait normally migrate up-river. The strict control over net sizes and rules against blocking the river to channel the fish into the net permit sufficient quantity of whitebait to reach the adult habitat and maintain stock levels. The whitebait themselves are very sensitive to objects in the river and are adept at dodging the nets.

The New Zealand whitebait is small, sweet and tender with a delicate taste that is easily overpowered if mixed with stronger ingredients when cooked. The most popular way of cooking whitebait in New Zealand is the whitebait fritter, which is essentially an omelette containing whitebait. Purists use only the egg white in order to minimise interfering with the taste of the bait. Foreigners frequently react with revulsion when shown uncooked whitebait, which resembles slimy, translucent worms.

The combination of the fishing controls, a limited season and the depletion of habitat as a result of forest felling during the era of colonisation results in limited quantities being available on the market. Whitebait is very much a delicacy and commands high prices to the extent that it is the most costly fish on the market, if available. It is normally sold fresh in small quantities, although some is frozen to extend the sale period. Nevertheless, whitebait can normally only be purchased during or close to the netting season.

Australian Whitebait;

In Australia whitebait refers to the juvenile stage of several predominantly galaxias species during their return to freshwater from the marine phase of their life cycle. Comments for New Zealand are generally applicable.

Species referred to as whitebait in Australia include Common galaxias *G. maculatus*, Climbing galaxias or koaro, *G. brevipinnis*, Spotted galaxias *G. truttaceus*, Tasmanian whitebait *Lovettia sealii*, Tasmanian mudfish *Neochanna cleaveri*, and Tasmanian smelt *Retropinna tasmanica*.

Whitebait were once subject to a substantial commercial fishery but today only recreational fishers are permitted to gather them, under strict conditions and for a limited season.

United Kingdom Whitebait

In the United Kingdom whitebait refers to young sprats, most commonly herring. They are normally deep-fried, coated in flour or a light batter, and served very hot with bread and butter.

Chinese Whitebait

Chinese whitebait is raised in fish farms and plentiful quantities are produced for export. The Chinese whitebait is larger than the New Zealand whitebait and not nearly so delicate. The frozen product is commonly available in food stores and supermarkets at reasonable prices.

White Croaker

White croaker (*Genyonemus lineatus*) is a species of croaker occurring in the Eastern Pacific. White croakers have been taken from Magdalena Bay, Baja California, to Vancouver Island, British Columbia, but are not abundant north of San Francisco. White croakers swim in loose schools at or near the bottom of sandy areas. Sometimes they aggregate in the surf zone or in shallow bays and lagoons. Most of the time they are found in offshore areas at depths of 3 to 30 metre (10 to 100 feet). On rare occasions they are fairly abundant at depths as great as 200 metre (600 feet).

The white croaker is the only species of in the genus *Genyonemus*. Other common names for the fish include Pasadena trout, tommy croaker, and little bass.

Description

The body of the white croaker is elongate and somewhat compressed. The head is oblong and bluntly rounded, with a mouth that is somewhat underneath the head. The colour is incandescent brownish to yellowish on the back becoming silvery below. The fins are yellow to white. The white croaker is one of five California croakers that have mouths located under their heads (subterminal). They can be distinguished from the California corbina and yellowfin croaker by the absence of a single fleshy projection, or barbel, at the tip of the lower jaw. The 12 to 15 spines in the first dorsal fin serve to distinguish white croakers from all the other croakers with subterminal mouths, since none of these has more than 11 spines in this fin.

White croakers eat a variety of fishes, squid, shrimp, octopus, worms, small crabs, clams and other items, either living or dead. While the ages of white croakers have not been determined conclusively, it is thought that some live as long as 15 or more years. Some spawn for the first time when they are between 2 and 3 years old. At this age they are only 12 to 15 cm (5 to 6 inches) long and weigh less than 45 gram (0.10 pounds). The largest recorded specimen was 41.4 cm (16.3 inches), no weight recorded; however, a 36.8 cm (14.5 inch) white croaker weighted 640 gram (1.41 pounds).

Fishing Information

These fish can be caught on almost any kind of animal bait that is fished from piers or jetties in sandy or sandy mud areas. In fact, they are so easily hooked that most anglers consider them a nuisance of the worst sort.

If a person desires to fish specifically for white croakers a tough, difficult-to- steal bait, such as squid, is recommended. When hooked, they put up little or no fight. Fishing and catching is good throughout the year.

Great White Shark

The great white shark, *Carcharodon carcharias*, also known as white pointer, white shark, or white death, is an exceptionally large lamniforme shark found in coastal surface waters in all major oceans. Reaching lengths of about 6 metres (20 ft) and weighing almost 2,000 kilograms (4,400 lb), the great white shark is the world's largest known predatory fish. It is the only known surviving species of its genus, *Carcharodon*. They are also regarded as an apex predator with its only real threats from humans and at least in one incident the orca. Although their diets overlap greatly, there are few reports of encounters between orcas and great whites, and they don't seem to directly compete with each other.

Distribution and Habitat

Great white sharks live in almost all coastal and offshore waters which have a water temperature of between 12 and 30° C (54° to 75° F), with greater concentrations off the southern coasts of Australia, off South Africa, California, Mexico's Isla Guadalupe and to a degree in the Central Mediterranean and Adriatic Seas. One of the densest known populations is found around Dyer Island, South Africa where much research on the shark is conducted. It can be also found in tropical waters like those of the Caribbean and has been recorded off Mauritius. It is a pelagic fish, but recorded or observed mostly in coastal waters in the presence of rich game like fur seals, sealions, cetaceans, other sharks and large bony fish species. It is considered an open-ocean dweller and is recorded from the surface down to depths of 1,280 metres (4,200 ft), but is most often found close to the surface.

In a recent study great white sharks from California were shown to migrate to an area between Baja California and Hawaii, where they spend at least 100 days of the year before they migrate back to Baja. On the journey out, they swim slowly and dive to up to 900 metres (3,000 ft). After they arrive, they change behaviour and do short dives to about 300 m (1,000 ft) for up to 10 minutes. It is still unknown why they migrate and what they do there; it might be seasonal feeding or possibly a mating area.

In a similar study a great white shark from South Africa was tracked swimming to the northwestern coast of Australia and back to the same location in South Africa, a journey of 20,000 kilometres in under 9 months.

Anatomy and Appearance

The great white shark has a robust large conical-shaped snout. It has almost the same size upper and lower lobes on the tail fin (like most mackerel sharks, but unlike most other sharks).

Great white sharks have a white belly and a grey back. The colouration makes it difficult for prey to spot the shark because it breaks up the shark's outline when seen from a lateral perspective. When viewed from above, the darker shade blends in with the sea.

Great white sharks, like many other sharks, have rows of teeth behind the main ones, allowing any that break off to be rapidly replaced. Their teeth are unattached to the jaw and are retractable, like a cat's claws, moving into place when the jaw is opened. Their teeth also rotate on their own axis (outward when the jaw is opened, inward when closed).

The teeth are linked to pressure and tension-sensing nerve cells. This arrangement seems to give their teeth high tactile sensitivity. A great white shark's teeth are serrated and when the shark bites it will shake its head side to side and the teeth will act as a saw and tear off large chunks of flesh.

Great white sharks often swallow their own broken off teeth along with chunks of their prey's flesh. These teeth frequently cause damage to the great white shark's digestive tract. However great white sharks often feed on stingrays and swallow the 'sting' as well, the barbed sting often getting stuck in the shark's intestines.

Size

The average length of a full-grown great white shark is 4 to 4.8 metres (13.3 to 15.8 ft), with a weight of 680 to 1,100 kilograms (1,500 to 2,450 lbs), females generally being larger than males. But the question of the maximum size of a great white shark has been subject to much debate, conjecture, and misinformation. Richard Ellis and John E. McCosker, both academic shark experts, devote a full chapter in their book, *The Great White Shark* (1991), to analysing various accounts of extreme size. Today, most experts contend that the great white shark's "normal" maximum size is about 6 metres (20 ft), with a maximum weight of about 1,900 kilograms (4,200 lb).

For some decades many ichthyological works, as well as the Guinness Book of World Records, listed two great white sharks as the largest individuals caught: an 11 metre (36 ft) great white captured in south Australian waters near Port Fairy in the 1870s, and an 11.3 metre (37.6 ft) shark trapped in a herring weir in New Brunswick, Canada in the 1930s. While this was the commonly accepted maximum size, reports of 7.5 to 10 metre (25 to 33.3 ft) great white sharks were common and often deemed credible.

Some researchers questioned the reliability of both measurements, noting they were much larger than any other accurately-reported great white shark. The New Brunswick shark may have been a wrongly-identified basking shark, as both sharks have similar body shapes. The question of the Port Fairy shark was settled in the 1970s, when J.E. Reynolds examined the shark's jaws and "found that the Port Fairy shark was of the order of 5 m (17 feet) in length and suggested that a mistake had been made in the original record, in 1870, of the shark's length.

Ellis and McCosker write that "the largest white sharks accurately measured range between 19 and 21 ft [about 5.8 to 6.4 m], and there are some questionable 23-footers [about 7 m] in the popular — but not the scientific — literature". Furthermore, they add that "these giants seem to disappear when a responsible observer approaches with a tape measure."

The largest specimen Ellis and McCosker endorse as reliably measured was 6.4 metres (21.3 ft) long, caught in Cuban waters in 1945 (though confident in their opinion, Ellis and McCosker note, however, that other experts have argued this individual might have been a few feet shorter).

Click here to see a photo of this Cuban shark. There are more photos available to the public of this particular Cuban specimen. The photos and the story of the shark hunt were published in the Polk Voice article "A Shark to Remember: The Story of a Great White Shark" by writer Eduardo J. Echenique.

There have since been claims of larger great white sharks, but, as Ellis and McCosker note, verification is often lacking and these extraordinarily large great white sharks have, upon examination, all proved of average size. For example, a female said to be 7.13 metres (over 23 ft) was fished in Malta in 1987 by Alfredo Cutajar. In their book, Ellis and McCosker agree this shark seemed to be larger than average, but they did not endorse the measurement. In the years since, experts eventually found reason to doubt the claim, due in no small part to conflicting accounts offered by Cutajar and others. A BBC photo analyst concluded that even "allowing for error ... the shark is concluded to be in the 18.3 ft [5.5 m] range and in no way approaches the 23 ft [7 m] reported by Abela."

According to the Canadian Shark Research Centre, the largest accurately measured great white shark was a female caught in August 1988 at Prince Edward Island off the Canadian (North Atlantic) coast and measured 6.1 metres (20.3 ft). The shark was caught by David McKendrick, a local resident from Alberton, West Prince.

The question of maximum weight is complicated by an unresolved question: when weighing a great white shark, does one account for the weight of the shark's recent meals. With a single bite, a great white can take in up to 14 kilograms (30 lb) of flesh, and can gorge on several hundred kilograms or pounds of food.

Ellis and McCosker write in regards to modern great white sharks that "it is likely that [great white] sharks can weigh as much as 2 tons", but also note that the largest recent scientifically measured examples weigh in at about 2 tonnes (1.75 short tons).

The largest great white shark recognised by the International Game Fish Association (IGFA) is one landed by Alf Dean in squth Australian waters in 1959, weighing 1,208 kilograms (2,664 lb). Several larger great white sharks caught by anglers have since been verified, but were later disallowed from formal recognition by IGFA monitors for rules violations.

Adaptations

Great white sharks, like all other sharks, have an extra sense given by the Ampullae of Lorenzini, which enables them to detect the electromagnetic field emitted by the movement of living animals. Every time a living creature moves it generates an electrical field and great whites are so sensitive they can detect half a billionth of a volt. This is equivalent to detecting a flashlight battery from 1,600 kilometres (1,000 miles) away.

To more successfully hunt fast moving and agile prey such as sea lions, the poikilothermic great white shark has developed adaptations that allow it to maintain a body temperature warmer than the surrounding water. One of these adaptations is a "rete mirabile" (Latin for "wonderful net"). This close web like structure of veins and arteries, located along each lateral side of the shark conserves heat by warming the cooler arterial blood with the venous blood that has been warmed by the working muscles.

This keeps certain parts of the body running at temperatures up to 14° C above the surrounding water, while the heart and gills remain at sea-temperature. When conserving energy (a great white shark can go weeks between meals), the core body temperature can drop to match the surroundings. A great white shark's success in raising its core temperature is an example of gigantothermy. Therefore, the great white shark can be considered an endothermic poikilotherm, because its body temperature is not constant but is internally regulated.

Diet

Great white sharks primarily eat fish, smaller sharks, turtles, dolphins, whale carcasses and pinnipeds such as seals and sea lions. Great whites have also been known to eat objects that they are unable to digest. In great white sharks above 3.41 metres (11 ft, 2 in) a diet consisting of a higher proportion of mammals has been observed.

Behaviour

A great white shark primarily uses its extra senses (i.e. electrosense and mechanosense) to locate prey from far off. Then, the shark uses smell and hearing to further verify that its target is food. At close range, the shark utilises sight for the attack.

Great white sharks' reputation as ferocious predators is well-earned, yet they are not (as was once believed) indiscriminate "eating machines". They typically hunt using an "ambush" technique, taking their prey by surprise from below.

Off Seal Island in South Africa studies have shown that the shark attacks most often in the morning, within 2 hours after sunrise. The reason for this is that it is hard to see a shark close to the bottom at this time. The success rate of attacks on average is 55 per cent in the first 2 hours, it falls to 40 per cent in late morning and after that the sharks stop hunting.

The ambush tactic, combined with the shark's ability to attain high speeds and its considerable mass, often cause great white sharks to breach (in a similar fashion to a whale breach) when attacking seals. Other sharks which have been observed to breach the water are the thresher shark, shortfin mako, longfin mako, spinner shark, basking shark, blacktip shark, salmon shark, porbeagle shark and the copper shark.

The great white shark will often deliver a massive disabling bite and then back off to allow the prey to expire. This tactic allows the animal to avoid combat with dangerous prey, such as sea lions. It also has allowed occasional rescue of humans bitten by the animal, though it appears to attack humans mostly in error.

The great white shark is one of the only sharks known to regularly lift its head above the sea surface to gaze at other objects such as prey; this is known as "spy-hopping".

This behaviour has also been seen in at least one group of blacktip reef sharks, but this might be a behaviour learned from interaction with humans (it is theorised that the shark may also be able to smell better this way, because smells travel through air faster than through water). They are very curious animals, and can display a high degree of intelligence and personality when conditions permit (such as in the clear waters off of Isla Guadalupe, Mexico).

Reproduction

There is still a great deal that is unknown about great white shark behaviour, such as their mating habits. Birth has never been observed, but several pregnant females have been examined. Great white sharks are ovoviviparous, the eggs developing in the female's uterus, hatching there and continuing to develop until they are born, at which point they are perfectly capable predators. The embryos can feed off unfecundated eggs. The delivery takes place in the period transitioning spring and summer.

The young, which number 8 or 9 (with a maximum of perhaps 14) for a single delivery, are about 1.5 metres (5 ft) long when born. Their teeth are provided with small side cusps. They grow rapidly, reaching 2 metres of length in the first year of life.

Almost nothing, however, is known about how and where the great white mates. There is some evidence that points to the near-soporific effect resulting from a large feast (such as a whale carcass) possibly inducing mating.

A great white shark can reproduce when a male's length is around 3.8 metres (12 ft) and a female's length is around 4 to 4.8 metres (13.3 to 15.8 ft). Their lifespan has not been definitively established, though many sources estimate 30 to 40 years. It would not be unreasonable to expect such a large marine animal to live longer however.

Relationship to Humans

Shark Attacks: More than any documented attack, Steven Spielberg's 1975 film *Jaws* provided the great white shark with the image of a "man eater" in the public mind. While great white sharks have been responsible for fatalities in humans, they typically do not target humans as prey: for example, in the Mediterranean Sea there were 31 confirmed attacks against humans in the last two centuries, only a small number of them deadly. Many incidents seem to be caused by the animals "test-biting" out of curiosity. Great white sharks are known to perform test-biting with buoys, flotsam, and other unfamiliar objects as well, and might grab a human or a surfboard with their mouth (their only tactile organ) in order to determine what kind of object it might be.

Other incidents seem to be cases of mistaken identity, in

which a shark ambushes a bather or surfer, usually from below, believing the silhouette it sees on the surface is a seal. Many attacks occur in waters with low visibility, or other situations in which the shark's senses are impaired. It has been speculated that the species typically does not like the taste of humans, or at least that the taste is unfamiliar.

However, some researchers have hypothesised that the reason the proportion of fatalities is low is not because sharks do not like human flesh, but because humans are often able to get out of the water after the shark's first bite. In the 1980s John McCosker noted that divers who dived solo and were attacked by great whites were generally at least partially consumed, while divers who followed the buddy system were normally pulled out of the water by their colleagues before the shark could finish its attack. Tricas and McCosker suggest that a standard attack modus operandi for great whites is to make an initial devastating attack on its prey, and then wait for the prey to weaken before going in to consume the ailing animal. A human's ability to get to land (or onto a boat) with the help of others is unusual for a great white's prey, and thus the attack is foiled.

Humans, in any case, are not healthy for great white sharks to eat because the sharks' digestion is too slow to cope with the human body's high ratio of bone to muscle and fat. Accordingly, in most recorded attacks, great whites have broken off contact after the first bite. Fatalities are usually caused by loss of blood from the initial limb injury rather than from critical organ loss or from whole consumption.

Biologist Douglas Long and Tyler B. write that the great white shark's "role as a menace is exaggerated; more people are killed in the US each year by dogs than have been killed by great white sharks in the last 100 years." However, such comments should be taken in context; interaction between humans and canines takes place far more regularly and in greater numbers than it does between humans and sharks.

Many "shark repellents" have been tested, some using

scent, others using protective clothing, but to date the most effective is an electronic beacon (POD) worn by the diver/ surfer that creates an electric field which disturbs the shark's sensitive electro-receptive sense organs, the ampullae of Lorenzini.

Great White Sharks in Captivity

All attempts to keep a great white shark in captivity prior tc August 1981 lasted 11 days or less. However, that month a great white broke previous records by lasting 16 days in captivity at SeaWorld San Diego before being released into the wild.

In 1984, shortly before opening day, the Monterey Bay Aquarium in Monterey, California housed its first great white shark, which died after 10 days. In July 2003, Monterey researchers captured a small female and kept it in a large, netted pen off Malibu for five days, where they had the rare success of getting the shark to feed in captivity before it was released. It was not until September 2004 that the aquarium was the first to place a great white on long-term exhibit. The young female, who was caught off the coast of Ventura, was kept in the aquarium's massive 1 million-gallon (3,800,000 litres) Outer Bay exhibit for 198 days before her successful release back to the wild in March 2005. She was tracked for 30 days after her early morning release. On the evening of August 31, 2006 the aquarium introduced a second shark to the Outer Bay exhibit. The juvenile male, caught outside Santa Monica Bay on August 17 , had its first official meal in captivity (a large salmon steak) on September 8, 2006 and as of that date, the shark was estimated to be 1.72 metres (5 ft 8 in) and to weigh approximately 47 kilograms (104 lb). He was released on January 16, 2007 after 137 days in captivity.

Probably the most famous great white shark to be kept in captivity was a female named "Sandy", which in August 1980 became the first and only great white shark to be housed at the Steinhart Aquarium in San Francisco, California. She was returned to the wild because she would not eat anything given to her and constantly bumped against the walls.

Shark Tourism and Cage Diving

Viewing sharks from the safety of a cage gives tourists an adrenaline rush and has become a booming industry. Common practice is to chum the water to draw in sharks for the tourists to view. These practices have raised the fear that as a result of this form of tourism, sharks are becoming accustomed to people in their environment and beginning to associate human activity with food - a potentially dangerous situation. It is claimed that certain methods of chumming where the bait on a wire is drawn towards the divers in the cage, sometimes resulting in the shark striking the cage, exacerbate this problem. Other operators purposefully draw the bait away from the cage causing the shark to swim past the divers.

Shark cage-diving is when a group of tourists, or those who wish to study the sharks up close are lowered into the water beside a boat, protected by a steel cage. From this view, point it is easier to view the sharks up close without the dangers (being bitten). Cage diving is most common off the coasts of Australia, South Africa, and Guadalupe Island off the coast of Baja California as it is here where great white sharks are most likely to be seen.

Companies respond that they are being made the scapegoats, as people try to find someone to blame for shark attacks on humans. Most point out that lightning tends to strike humans more often than sharks bite humans. Their position is that further research needs to be done before banning practices such as chumming which are said to alter sharks natural behaviour.

This section is a stub. You can help by expanding it.

Conservation Status

It is unclear how much a concurrent increase in fishing for great white sharks had to do with the decline of great white shark population from the 1970s to the present. No accurate numbers on population are available, but populations have clearly declined to a point at which the great white shark is now considered endangered. Their reproduction is slow, with sexual maturity occurring at about nine years of age, the population, therefore, can take a long time to rise.

In 2005, a tagged great white shark named "Nicole" was recorded swimming from South Africa to Australia and back, a 22,000 kilometre round trip. Researchers believe it may have undertaken this journey to mate, and hope studies such as this will produce more effective conservation measures.

The Convention on International Trade in Endangered Species (C.I.T.E.S.) has put the great white shark on its 'Appendix II' list of endangered species.

The shark is targeted by fishermen for its jaws, teeth, and fins, and as a game fish. The great white shark, however, is rarely an object of commercial fishing, although its flesh is considered valuable. If casually captured (it happens for example in some tonnare in the Mediterranean), it is sold as *smooth-hound shark*.

From April 2007, great white sharks will be fully protected within 200 nautical miles of New Zealand and additionally from fishing by New Zealand-flagged boats outside this range.

Related Species

Dental features and the extreme size of both the great white shark and the megalodon, *Carcharodon megalodon*, lead some scientists to believe they were closely related, however there is much doubt about this hypothesis and other scientists would place the megalodon and white shark as distant relatives-sharing the family Lamnidae but no closer relationship. Megalodon is only known from its teeth, and may have reached sizes of 12 metres (40 ft) or more, considerably larger than even the largest great white sharks.

From time to time it is suggested that megalodon might still exist. Megalodon teeth have been found from as recently as 10,000 to 12,000 years ago, though some have questioned the reliability of these estimates. However, while megalodon fossils are widespread and plentiful, no evidence has surfaced that the species is anything but extinct.

Other evidence suggests that the great white shark is more closely related to the mako shark than to the megalodon.

Whitetip Reef Shark

The whitetip reef shark, *Triaenodon obesus*, is a requiem shark of the family Carcharhinidae, the only member of the genus *Triaenodon*.

Habitat and Distribution

The whitetip reef shark is one of the most common sharks found in shallow tropical and warm temperate water around coral reefs in the Indian and Pacific oceans. It occurs at depths down to 330 m.

Anatomy and Appearance

As its name suggests, the tips of the shark's first dorsal fin and upper caudal fin are white. The upper body is grey/ brownish. Their average length is about 140 to 160 cm and the maximum reported length is 244 cm. Its head is broad and flat.

Diet

The whitetip reef shark feeds primarily on crustaceans, octopuses and fish.

Behaviour

This bottom dwelling shark is nocturnal and is often seen resting on the bottom during the day, sometimes in small groups. It is not aggressive and will generally swim away if disturbed, although it may bite if harassed. At night it hunts among crevices in the reef.

Reproduction

Reproduction is viviparous, with 1 to 5 pups in a litter, the gestation period being at least 5 months. The shark's size at birth ranges from 50 to 60 cm. It is estimated that this shark can live for about 25 years and it reaches maturity after about 5 years.

Whiting

Many types of fish have been given the common name whiting. The fish originally known by that name in English is *Merlangius merlangus*, in the family Gadidae. In the USA it is known as the English whiting.

In the USA, the name whiting on its own is often used for various species of hake, genus *Merluccius*. In Canada, it is used for the Alaska pollock *Theragra chalcogramma*. Like the true whitings, these are all members of the order Gadiformes. In India and Australia, the name whiting (though usually with an adjective before it) is used for various species of the genus *Sillago*, in the smelt-whiting family Sillaginidae, which is in the order Perciformes.

The whiting *Merlangius merlangus* is an important food fish in the eastern North Atlantic, northern Mediterranean, western Baltic, and Black Sea. Until the later twentieth century, it was a cheap fish, regarded as food for the poor or for pets, but the general decline in fish stocks means that it is now more highly valued. The other fish that have been given the name whiting are mostly also food fish.

Wobbegong

Wobbegong is the common name given to the eight species of carpet sharks in the family Orectolobidae. They are found in shallow temperate and tropical waters of the western Pacific Ocean and eastern Indian Ocean, chiefly around Australia and Indonesia, although one species (the Japanese wobbegong, *Orectolobus japonicus*) occurs as far north as Japan.

Wobbegongs are bottom-dwelling sharks which spend much of their time resting on the sea floor, often among rocks or under ledges. The largest species, the spotted wobbegong, *Orectolobus maculatus*, grows up to 3.2 m long. Wobbegongs are well camouflaged with a symmetrical pattern of bold markings which resembles carpet. Because of this striking pattern, wobbegongs and their close relatives are often referred to as carpet sharks. The camouflage is improved by the presence of small vegetation-like flaps of skin around the wobbegong's mouth. Wobbegongs make use of their relative invisibility to hide among rocks and ambush smaller fish which swim too close (animals which feed in this way are called *ambush predators*).

Wobbegongs are generally not dangerous unless they are provoked. They have bitten people who accidentally step on them in shallow water; they may also bite scuba divers or snorkellers who poke or handle them, or who block their escape route.

Wobbegongs are very flexible and can easily bite a hand that is holding on to their tail. They have many small but sharp teeth and their bite can be severe, even through a wetsuit; having once bitten, they have been known to hang on and be very difficult to remove. To avoid being bitten, do not harass or encircle a wobbegong when diving, and pay close attention near the bottom to avoid accidental contact with a wobbegong you have not seen.

Although wobbegongs do not eat humans, humans frequently eat wobbegongs; the flesh of a wobbegong or other shark is called flake and it is often used in fish and chips in Australia. Wobbegong skin is also used to make leather.

The word *wobbegong* is believed to come from an Australian Aboriginal language.

Genera and Species

- *Orectolobus*
  - — *Orectolobus halei* (Whitley, 1940).
  - — *Orectolobus hutchinsi* (Last, Chidlow & Compagno, 2006).
  - — Tasselled wobbegong, *Orectolobus dasypogon* (Bleeker, 1867)
  - — Japanese wobbegong, *Orectolobus japonicus* (Regan, 1906)
  - — Spotted wobbegong, *Orectolobus maculatus* (Bonnaterre, 1788)
  - — Ornate wobbegong, *Orectolobus ornatus* (De Vis, 1883)

— Northern wobbegong, *Orectolobus wardi* (Whitley, 1939)

— Western wobbegong, *Orectolobus sp. A*

- *Sutorectus*

— Cobbler wobbegong, *Sutorectus tentaculatus* (Peters, 1864)

Wolf Herring

The wolf herrings are a family (Chirocentridae) of two marine species of ray-finned fish related to the herrings.

Both species have elongated bodies and jaws with long sharp teeth that facilitate their ravenous appetites, mostly for other fish. Both species reach a length of 1 m. They have silvery sides and bluish backs.

They are commercially fished, and marketed fresh or frozen.

Species

- Dorab wolf-herring, *Chirocentrus dorab,* is found in warm coastal waters from the Red Sea to Japan and Australia.
- Whitefin wolf-herring, *Chirocentrus nudus,* is found in a similar range, and is difficult to distinguish from C. *dorab* (the former has a black mark on its dorsal fin). This species is also known to eat crabs in addition to its usual diet of smaller fish.

Wormfish

Wormfishes are a family, Microdesmidae, of goby-like fishes in the order Perciformes.

They are found in shallow tropical waters, both marine and brackish, often burrowing in estuarine mud. They are small fishes, the largest species reaching only about 12 cm in length.

Species

The family formerly included five genera of dartfishes in the subfamily Ptereleotrinae, but this inclusion made the family polyphyletic and these fishes are now grouped in their own family, Ptereleotridae.

FishBase lists 27 species in five genera:

- Genus *Cerdale*
  - — *Cerdale fasciata* (Dawson, 1974).
  - — Pugjaw wormfish, *Cerdale floridana* (Longley, 1934).
  - — Spotted worm goby, *Cerdale ionthas* (Jordan & Gilbert, 1882).
  - — *Cerdale paludicola* (Dawson, 1974).
  - — *Cerdale prolata* (Dawson, 1974).
- Genus *Clarkichthys*
  - — Flagtail wormfish, *Clarkichthys bilineatus* (Clark, 1936).
- Genus *Gunnellichthys*
  - — *Gunnellichthys copleyi* (Smith, 1951).
  - — Curious wormfish, *Gunnellichthys curiosus* (Dawson, 1968).
  - — *Gunnellichthys grandoculis* (Kendall & Goldsborough, 1911).
  - — *Gunnellichthys irideus* (Smith, 1958).
  - — Onespot wormfish, *Gunnellichthys monostigma* (Smith, 1958).
  - — Onestripe wormfish, *Gunnellichthys pleurotaenia* (Bleeker, 1858).
  - — Yellowstripe wormfish, *Gunnellichthys viridescens* (Dawson, 1968).
- Genus *Microdesmus*
  - — *Microdesmus aethiopicus* (Chabanaud, 1927).
  - — Olivaceous wormfish, *Microdesmus affinis* (Meek & Hildebrand, 1928).
  - — *Microdesmus africanus* (Dawson, 1979).
  - — *Microdesmus bahianus* (Dawson, 1973).
  - — Stippled wormfish, *Microdesmus carri* (Gilbert, 1966).
  - — Banded worm goby, *Microdesmus dipus* (Gunther, 1864).
  - — Spotback wormfish, *Microdesmus dorsipunctatus* (Dawson, 1968).

— Lancetail wormfish, *Microdesmus lanceolatus* (Dawson, 1962).

— Pink wormfish, *Microdesmus longipinnis* (Weymouth, 1910).

— *Microdesmus luscus* (Dawson, 1977).

— Rearfin wormfish, *Microdesmus retropinnis* (Jordan & Gilbert, 1882).

— Spotside wormfish, *Microdesmus suttkusi* (Gilbert, 1966).

- Genus *Paragunnellichthys*

— Seychelle's wormfish, *Paragunnellichthys seychellensis* (Dawson, 1967).

— *Paragunnellichthys springeri* (Dawson, 1970).

Wrasse

The wrasses are a family, Labridae, of marine fish, many of which are brightly coloured. The family is large and diverse, with about 500 species in 60 genera.

Some species are popular aquarium fish; others are popular food fish. In the western Atlantic, the most commonly eaten is the tautog. In the eastern Atlantic, the term 'wrasse' by itself generally refers to the Ballan wrasse, *Labrus bergylta*.

Wrasses have protractile mouths, usually with separate jaw teeth that jut outwards. The dorsal fin has 8 – 21 spines and 6 – 21 soft rays, usually running most of the length of the back.

Some wrasses are widely known for their role as symbiotic fish, similar to the actions and those ascribed to the Egyptian plover: other fish will congregate at wrasse cleaning stations and wait for wrasses to swim into their open mouths and gill cavities to have gnathiid parasites removed. The cleaner wrasses are most well known to feed on dead tissue and scales and ectoparasites although they are also known to 'cheat' through the removal of healthy tissue and mucus which is costly for the client fish to produce. The bluestreak cleaner wrasse, *Labroides dimidiatus* is one of the most common cleaners found on tropical reefs. Few cleaner wrasses have been observed being eaten by predators, possibly because the removal of parasites from the predator fish is more important for the survival of the predator than the short-term gain of eating the cleaner.

Other species of wrasse, rather than having fixed cleaning stations, specialise in making "house calls" - that is, their "clientele" are those fish that are too territorial or shy to go to a cleaning station.

Aquaria

Some brightly-coloured wrasses are popular for salt-water aquaria. They are reef safe.

Wrymouth

The wrymouth, *Cryptacanthodes maculatus,* sometimes called ghostfish, is a slim, eel-like creature belongs to the wrymouth family Cryptacanthodidae. It outgrows the blennies, its relatives, and may reach 3 ft. A low spiny dorsal fin stands along the entire back. This includes about 70 spines and unites with the caudal and anal fins. Small eyes lie near the top of its big head. The mouth slants sharply above a ponderous lower jaw. Along its reddish brown upper sides extend several irregular rows of small dark spots. The dorsal and anal fins also show spots.

Distribution and Habitat

Dwelling along the North Atlantic coast of North America, it ordinarily hugs the bottom from shallow water down to considerable depths. It likes to burrow in the mud. About 2 inches around and 1.5 to 3 in below the surface, the burrow branches into tunnels. Sometimes, however, the burrow stands as high as 4 feet above the low water mark. It begins in a centrally placed mound with smaller openings along the tunnel and at its end. Usually it eats small invertebrates such as shrimp and crabs. Once in a while it grabs a small fish. It spawns in the winter, probably in deep water.

In the Aquarium

In aquaria the wrymouth has been known to utilise a rubber tube as a ready-made burrow.

Yellow-and-black triplefin

The yellow-and-black triplefin, *Forsterygion flavonigrum*, a triplefin of the genus *Forsterygion*, is found around the north of the North Island of New Zealand at depths of between 15 and 30 m, in reef areas of broken rock. Its length is between 4 to 7 cm.

Its non-breeding colouration is a pale pinkish head with a yellowish body and tail, with a black mask across the eyes which continues in a stripe down the centre of the body gradually changing to a darker yellow.

The breeding colouration is spectacular - the head and tail become black, while the rest of the body becomes bright yellow. Yellow-and-black triplefins guard their nest. After spawning the non-breeding colours rapidly return.

Yellowbanded Perch

The yellowbanded perch, *Acanthistius cinctus*, is a large marine fish of the grouper family, found around eastern Australia including Lord Howe Island, islands off the east coast of Northland in the North Island of New Zealand, and the Kermadec Islands, in caves and archways of rocky reefs. Their length is between 40 and 60 cm.

The yellowbanded perch is narrower bodied than other groupers, and its colouring of a series of pale yellow and dark green-black stripes make it impossible to confuse this fish with any other serranid.

Yellow Bass

The yellow bass or barfish, *Morone mississippiensis*, is a freshwater fish native to the south and midwestern United States. Though sometimes confused with white bass or striped bass, it is distinguished by its yellow belly and the broken pattern in its lowermost stripes. The species name "mississippiensis" refers to the Mississippi River, where it was first described and is still most commonly found. Yellow bass reach an average length of 28 cm (11 in).

Yellow Bittern

The Yellow Bittern (*Ixobrychus sinensis*) is a small bittern. It is of Old World origins, breeding in tropical Asia from India and Sri Lanka east to Japan and Indonesia. It is mainly resident, but some northern birds migrate short distances. There is a single record from Britain, from Radipole Lake, Dorset on November 23rd 1962 and one in Alaska as well—however, the BOU have always considered this occurrence to be of uncertain provenance and currently it is not accepted as a genuine wild vagrant.

This is a small species at 38cm length, with a short neck and longish bill. The male is uniformly dull yellow above and buff below. The head and neck are chestnut, with a black crown. The female's crown, neck and breast are streaked brown, and the juvenile is like the female but heavily streaked brown below, and mottled with buff above.

Their breeding habitat is reedbeds. They nest on platforms of reeds in shrubs. 4-6 eggs are laid. They can be difficult to see, given their skulking lifestyle and reedbed habitat, but tend to fly fairly frequently, when the striking contrast between the black flight feathers and the other wise yellowish plumage makes them unmistakable.

Yellow Bitterns feed on insects, fish and amphibians.

Protected Status

The Yellow Bittern (Ixobrychus sinensis) is protected under the Migratory Bird Treaty Act of 1918.

Yellow-edged Moray

The yellow-edged moray, *Gymnothorax flavimarginatus*, is a moray eel of the family Muraenidae, found in the Indo-Pacific oceans from the Red Sea and South Africa eastward to the Tuamotus and Austral islands, north to the Ryukyu and Hawaiian islands, south to New Caledonia, and in the eastern Pacific from Costa Rica, Panama and the Galapagos Islands, at depths down to 150 m. Its length is up to 240 cm.

The yellow-edged moray is found along drop-offs in coral or rocky areas of reef flats and protected shorelines to seaward reefs. It feeds on cephalopods, fishes, and crustaceans.

It most often appears on the reef after a fish has been speared during daylight. The regularity and promptness of such appearances make it clear that the yellow-edged moray is especially sensitive to stimuli emanating from an injured or stressed fish. It is eaten in some parts of the Indo-Pacific.

Colouration is yellowish, densely mottled with dark brown, with the front of the head purplish grey and the posterior margins of fins yellow-green. The eyes are reddish and the gill opening is a black blotch. Juveniles sometimes are bright yellow with brown blotches.

Yellow-eye Mullet

The yellow-eye mullet, *Aldrichetta forsteri*, is a mullet of the family Mugilidae, the only species in the genus *Aldrichetta*. It is found around New Zealand, the Chatham Islands, and southern Australia, from the surface to depths of 50 m. Its length is between 20 and 40 cm. The yellow-eye mullet is similar to the flathead mullet but has a more pointed snout and does not grow to such a large size. It has a larger mouth than the flathead mullet and the teeth are larger and more numerous.

The yellow-eye mullet schools in large numbers in summer and enters bays and estuaries, but does not usually enter fresh water. The back is olive-green, and silver, usually with a yellow tinge, on the belly. As its common name suggests the eyes are a distinctive bright yellow.

Historic Nominal Species

Aldrichetta forsteri nonpilcharda: Whitley, 1951; *Mugil forsteri*: Valenciennes, 1836; *Mugil albula*: Forster, 1801; *Agonostoma diemensis*: (Richardson, 1840); *Dajaus diemensis*: Richardson, 1840.

Yellowhead Jawfish

Yellow-headed jawfish, *Opistognathus aurifrons*, live in coral reefs. They use their mouths like scoops to dig burrows in the sand. Males also use their mouths to carry eggs until they hatch.

These fish have a wonderful building propensity. They are often seen carrying (in their mouths) pieces of shell from one location to another, or from out of their burrows, and placing them in more preferred locations.

Yellowfin Croaker

Yellowfin croacker (*Umbrina roncador*) is a species of croaker occurring from the Gulf of California, Mexico, to Point Conception, California. They frequent bays, channels, harbours and other nearshore waters over sandy bottoms. These croakers are more abundant along beaches during the summer months and may move to deeper water in winter.

Other common names include yellowfin drum, Catalina croaker, yellowtailed croaker, and golden croaker.

Description

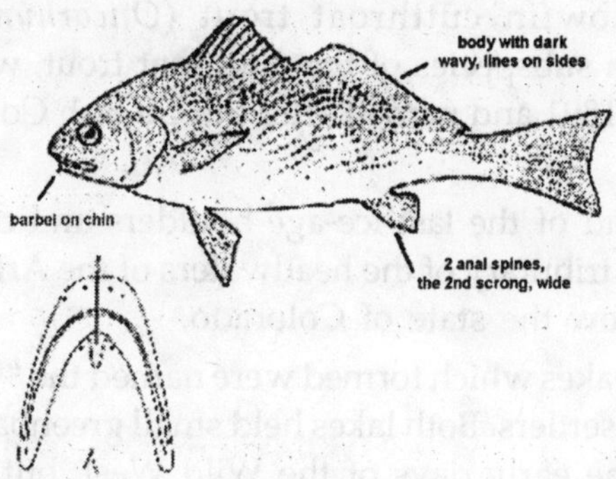

The body of the yellowfin croaker is elliptical-elongate with the back somewhat arched. The head is conical and blunt. The colour is iridescent blue to gray with brassy reflections on the back diffusing to silvery white below. The sides and back have many diagonal dark wavy lines. The fins are yellowish except for the dark dorsal fins. The yellowfin croaker differs from other California croakers in having a single fleshy projection, a barbel, on the lower jaw and two heavy spines at the front of the anal fin.

The diet of the yellowfin croaker consists mainly of small fishes and fish fry; however, invertebrates such as small crustaceans, worms and molluscs are also eaten in large numbers.

Spawning takes place during the summer months when this species is most common along the sandy beaches. Maturity is apparently not reached until the fish are slightly over 9 inches long. The largest recorded specimen was 20.13 inches; no weight reported. However, an 18-inch yellowfin croaker weighed 4.5 pounds.

Fishing Information

Yellowfin croaker are most often taken by surf anglers using softshelled sand crabs, worms, mussels, clams or cut fish as bait.

Yellowfin Cutthroat Trout

The yellowfin cutthroat trout (*Oncorhynchus clarki macdonaldi*), a subspecies of the cutthroat trout, was officially identified in 1891 and named after the US Fish Commissioner, MacDonald.

At the end of the last ice-age boulders and clay moraine blocked off a tributary of the headwaters of the Arkansas River in what is now the state of Colorado.

The two lakes which formed were named the "Twin Lakes" by the area's settlers. Both lakes held small greenback cutthroat trout from the early days of the Wild West, but in the mid-1880s reports circulated of much larger trout, up to 10lb in weight, with bright yellow fins.

In July 1889, Professor D. S. Jordan and G. R. Fisher visited Twin Lakes and published their discoveries in the 1891 Bulletin of the United States Fish Commission. They found both the

Anmol - 9-10-07
45783 5×5